WIND ENERGY IN ELECTRICITY MARKETS WITH HIGH WIND PENETRATION

WIND ENERGY IN ELECTRICITY MARKETS WITH HIGH WIND PENETRATION

JULIO USAOLA
AND
EDGARDO D. CASTRONUOVO

Nova Science Publishers, Inc.
New York

Copyright © 2009 by Nova Science Publishers, Inc.

All rights reserved. No part of this book may be reproduced, stored in a retrieval system or transmitted in any form or by any means: electronic, electrostatic, magnetic, tape, mechanical photocopying, recording or otherwise without the written permission of the Publisher.

For permission to use material from this book please contact us:
Telephone 631-231-7269; Fax 631-231-8175
Web Site: http://www.novapublishers.com

NOTICE TO THE READER

The Publisher has taken reasonable care in the preparation of this book, but makes no expressed or implied warranty of any kind and assumes no responsibility for any errors or omissions. No liability is assumed for incidental or consequential damages in connection with or arising out of information contained in this book. The Publisher shall not be liable for any special, consequential, or exemplary damages resulting, in whole or in part, from the readers' use of, or reliance upon, this material.

Independent verification should be sought for any data, advice or recommendations contained in this book. In addition, no responsibility is assumed by the publisher for any injury and/or damage to persons or property arising from any methods, products, instructions, ideas or otherwise contained in this publication.

This publication is designed to provide accurate and authoritative information with regard to the subject matter covered herein. It is sold with the clear understanding that the Publisher is not engaged in rendering legal or any other professional services. If legal or any other expert assistance is required, the services of a competent person should be sought. FROM A DECLARATION OF PARTICIPANTS JOINTLY ADOPTED BY A COMMITTEE OF THE AMERICAN BAR ASSOCIATION AND A COMMITTEE OF PUBLISHERS.

LIBRARY OF CONGRESS CATALOGING-IN-PUBLICATION DATA
Available upon request

ISBN: 978-1-60741-153-6

Published by Nova Science Publishers, Inc. ✢ *New York*

CONTENTS

Preface		vii
Chapter 1	Introduction	1
Chapter 2	Short Term Wind Power Prediction	3
Chapter 3	Bidding Strategies for Wind Farm Owners	23
Chapter 4	Optimal Distributed Management of Wind Power	43
Chapter 5	Conclusion	65
References		67
Index		71

PREFACE

The great amount of wind power recently connected to electric power systems makes necessary new grid analysis tools and control strategies. Some of these tools are short term wind power prediction programs, which have been recently developed and are already in use in many System Operators and wind power producers. Prediction programs also allow the participation of wind energy in the electricity market by keeping the economic losses due to power imbalances for wind generators within acceptable values, specially if the uncertainty of the predictions is evaluated and used in the bidding process. Such participation intends to make the power system operation easier. Distributed control of power system, and specially optimized control of clusters of wind generators is another measure that is being taken in order to maintain the levels of power system security and to minimize the consequences of possible and temporary wind power curtailments, due to grid security reasons, on the affected wind farms. The contents of this chapter include a detailed description of short term wind power prediction programs, an evaluation of the economic losses due to imbalances incurred by wind generators participating in an electricity market, and an optimization strategy aimed at minimizing the consequences of power curtailments.

Chapter 1

INTRODUCTION

In the last years, a great and sudden increase of wind generation has taken place in many countries. Most of its power injects their production into the power grids, which have to face the problems of a different form of generation, with peculiar features. Among them, perhaps the most annoying is the variability of the primary resource, the wind, and its difficulty in dispatching this power. It is well known that electric energy cannot be easily stored, and this means that wind power fluctuations must be balanced by other types of generation. This uncertainty stresses the system management and operation, and may add extra costs to the electricity.

These disadvantages are clearly compensated by the renewable nature of this source of energy, and for the null cost of the resource, that reduces the overall cost of electricity. However, the need for new analysis and control tools for the power grids is evident, if we expect to keep the reliability of these grids at a maximum level and to limit all avoidable cost to a minimum.

Short term wind power prediction programs are one of the tools that have been recently developed to help power system operation. They use the output of numerical weather forecast, and produce a reasonably accurate prediction for the wind power production of the next hours or few days. With this information, system operation is substantially helped.

Prediction tools also allow the participation of wind energy in electricity markets. Wind farm owners may commit themselves to produce the predicted wind power for the next hours. Because of the inaccuracy of these predictions, actual generation differs from the predicted values committed, and this imbalance generate an extra cost that must be paid by those who produce it, namely the wind generators. Nevertheless, this extra cost is reasonably small if a state-of-the-art prediction program is used, and may even be reduced with an adequate subsidy policy. The generators will be interested in reducing this imbalance cost as much as possible, and to fit the generation to the predicted value. In this incentive for adjusting the production lays the main benefit of the participation of wind generation in the markets – it is a way of reducing the need of balancing energy in a power system.

Another problem of wind generation is that older wind farms cannot withstand non standard grid conditions such as under/over voltage or under/over frequency. This means that much of wind generation disconnects if voltage goes below 90% of the rated value. Although this is not important when there are just a few wind farms connected to a big grid, when there is a substantial wind penetration, these disconnections may mean the switching off of hundreds on MW and may have a strong effect on power system stability. Most of the recent generation has been designed to remain connected under severe undervoltages, but a great part of wind farms cannot still deal with undervoltages produced by a short circuit, an incident quite common in the power systems. Modern system operation must tackle this problem by the use of new control strategies.

In the following sections, these problems have been addressed, describing new strategies and control tools that have, or are being implemented in power systems with high wind power penetration. First of all, short term wind power prediction tools are described, their typical performance presented and examples of actual prediction programs given in order to show their principles, possibilities and limitations.

Then, the economic consequences of participation of wind energy in the electricity markets are evaluated. Numerical tests show that economic losses of wind producers, due to imbalance costs, are assumable if they use a short term wind power prediction program. Knowledge and adequate use of the uncertainty of this power prediction may even reduce this imbalance cost.

Finally an optimized control strategy for distributed management of wind power is presented. This strategy minimize the consequences of possible power curtailments imposed by the System Operator and uses market mechanisms to allocate the power decrease to those wind farms more willing or capable to control their production. This strategy is also aimed at improving the manageability of power systems with a heavy wind power penetration.

Chapter 2

SHORT TERM WIND POWER PREDICTION

2.1. GENERAL BACKGROUND

Short term wind power prediction programs are tools that provide an estimation of the future power production of a wind farm, or a group of wind farms, in the next hours. For this purpose, they use meteorological forecasts coming from a Numerical Weather Prediction (NWP) tool, and sometimes real time SCADA data from the wind farms. Typically, their predictions are issued at least once a day, although many prediction programs refresh these predictions once per hour, or even with greater frequency. The output is the average hourly (typically) production forecasted for the prediction time horizon. These programs have the following main applications:

- They aid the power system operation by estimating the wind power production for the future hours. This helps to foresee the congestions in the system, to program the power reserve, or to prevent extreme events under adverse meteorological conditions. This is why System Operators of most countries with high wind penetration use them.
- They are also used by the balancing companies, i.e., those who have to buy energy for the customers connected to their grids, where there are also wind farms connected. An inaccurate prediction makes these companies to pay for their imbalances.
- They make possible the participation of wind power in electricity markets. Power predictions can be used to make bids in the market. Although there are deviations between the predicted power and the actual generation, prediction programs keep these imbalances small enough

(around 10% of the income) to make this participation profitable. This participation makes easier the integration of wind energy in the electric systems.
- They are also useful for wind farm maintenance scheduling, since they estimate future production. However, accuracy of short term wind power prediction tools decrease drastically after a few days, and for this reason, their application to this purpose is limited.

A short term wind power prediction tool usually has the outline shown in Figure 1. As mentioned, the basic inputs to the program are meteorological forecasts. Additionally, there may also be inputs from a SCADA system, with data of the actual production of one or several wind farms. This allows the regular updating of prediction, based on more recent data. With these fresh data some prediction tools are able to issue new predictions each hour, or even more frequently. SCADA data include power output from the wind farms most of the times, but it may include also meteorological data, such as measured wind speed and direction, temperature or atmospheric pressure. In any case, the real production data from the wind farms must be known, in order to check the accuracy of the predictions and to tune the models. In some prediction programs, this information is received regularly, but some time (one day, for instance) after real time.

It is also necessary to have data of the wind farms, such as rated power, type and availability of wind turbines, etc. Also information about the wind farm site is needed, as the UTM coordinates, and some prediction tools need information about terrain: roughness topography, turbine site etc.

The output of these programs is the average wind farm production for the next hours. The time scope of this prediction depends on the horizon of the numerical weather prediction available. Typically, predictions are issued for the next 48 hours, but longer time horizons are possible, sometimes at the price of a poorer resolution.

An important part of the wind power prediction program is the management of past predictions that are used to revise the modelling and to assess the accuracy the performance of the prediction tool.

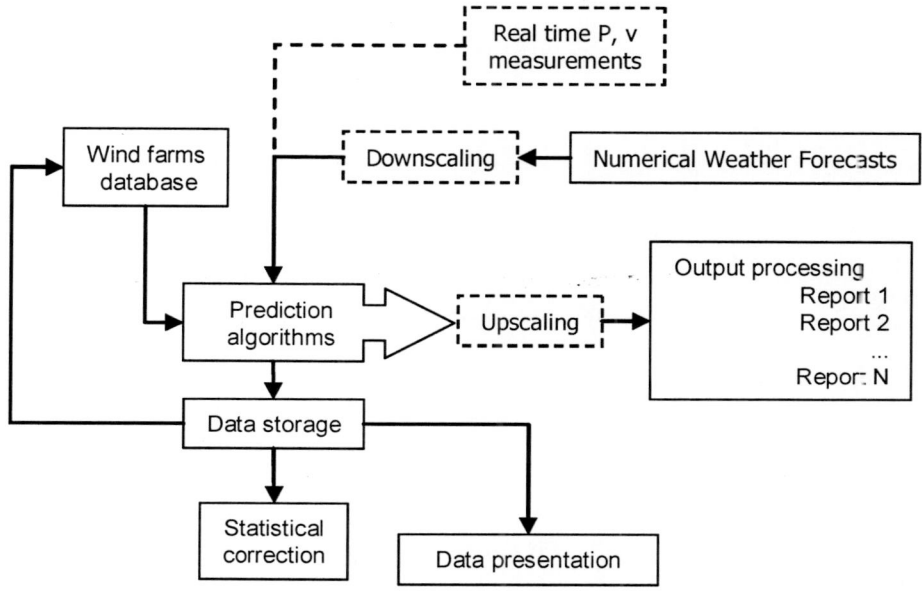

Figure 1. Elements of a short term wind power prediction program.

2.2. NUMERICAL WEATHER PREDICTIONS (NWP)

Numerical Weather Predictions (NWP) are the basic input to a prediction program. They are usually sent by a provider of this service, who typically are National Weather Services, or companies specialized in this field. New predictions are received only some times each day, due to the high computation requirements of the programs employed.

Different numerical models provide weather predictions around the world. For instance, the European Centre for Medium-Range Weather Forecasts (ECMWF) provides (among many other products) wind speed and direction predictions, for many places throughout the Earth, each six hours for up to 72 hours in advance at 3 hours intervals, and up to 240 hours in advance at 6 hours intervals, with a resolution of 0.25° x 0.25° lat/long. Predictions are produced for the whole world. This organization is funded by 28 countries, mostly European ones. There are other projects running that provide also predictions on a regional or global scale, as the project HIRLAM, Aladin, LokalModell, or the MRF models of the USA. Numerous references can be found in literature and elsewhere.

These forecasts have usually low resolution, since their results are extended to a wide area in a continental scale. This is why a downscaling procedure is often used. Downscaling uses meso – or microscale models to fit the wind speed predictions to the wind farms site. This is usually a difficult task, since it may require detailed modelling of the terrain, which for some farms is rather rugged, since they are placed to profit of local effects.

2.3. Prediction Methods

The core of a prediction program is the algorithm used to produce the predictions. All current prediction programs use at least one of two kinds of methods of prediction wind power – physical and statistical methods. Physical methods try to model the wind farm equations according to the aerodynamical behaviour of wind turbines in the actual site, and the local effects of wind speed and directions. Statistical methods try to reproduce the behaviour of the wind farm from past data under different conditions. There are different techniques to do this: time series analysis, nonparametric methods and Artificial Intelligence based methods have been successfully used.

The simplest prediction method is called *persistence*. This method consists of assuming that the future prediction, for all the time horizon considered, is the current production of the wind farm. Although this method has little practical value, it is considered as a lower threshold for the performance of a prediction method.

Predictions may be also easily made by taking the NWP (downscaled or not) and converting this value to power by means of a P-v curve. This curve could be made as easily as to take the sum of the standard power curves of the wind farms as shown in Figure 9 a), or it may be obtained from past power and wind measurements in the wind farm, as shown in Figure 9 b). It could also be an empirical curve that expresses the relation between past wind forecasts, and power measurements. The power curve of a wind farm is different according to the wind direction, due to wake effects or to a terrain with different roughness in different directions, so a more sophisticated model could be a "power surface" that takes into account this different behaviour. However, actual methods are more complex than these, because these simple methods do not take into account the different conditions of the wind farm along time, nor the characteristics of the site.

The physical models of most prediction programs, such as [19] or [21] receive the wind speed and direction data from a NWP, and adapt them to local characteristics. Then, this wind is transformed into output power using the wind farm power curve, which normally is a complex model. This output is modified according to the wind farm efficiency that accounts for the wake effects. A Model Output Statistics (MOS) module is normally used to improve the results. This module can be applied either to the incoming wind, or the outcoming power. MOS require data of the farm for several months, and detailed information of the turbine site and terrain. Physical models usually require long computation times, and for this reason their predictions are updated just few times a day.

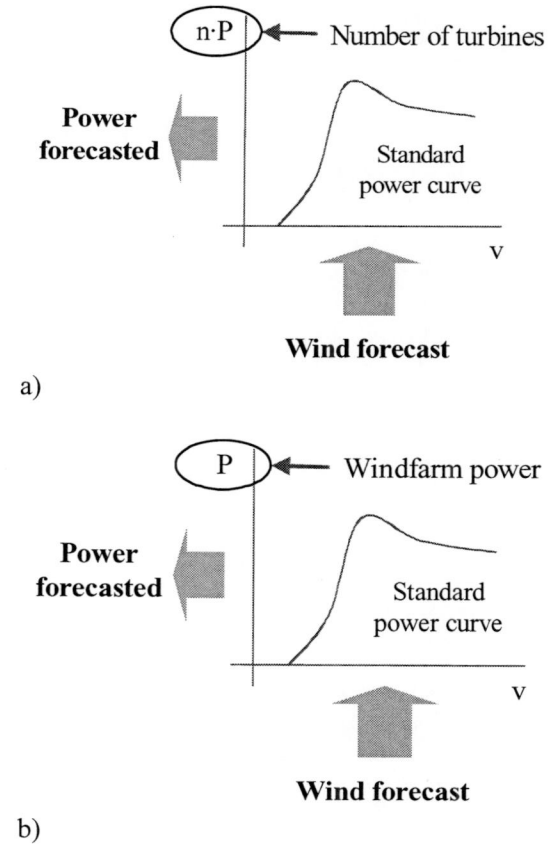

Figure 2. Simple prediction methods.

Statistical models also use the outcome of NWP programs, as well as real time power production (and sometimes other SCADA data from the wind farm), and the power predictions are obtained from a statistical module that models the wind farm from the inputs and outputs through a long period. Statistical prediction models usually do not have high computational requirements, and the predictions can be updated typically every hour or even more frequently. Different methods can be included in this family. For instance [19] uses time series models with exogenous inputs (the NWP), with parameters that vary adaptatively with time. It also models the power curve of the wind farm according to past data and recent conditions. Other models use fuzzy artificial neural networks for this purpose [27]. Statistical models need also to work for some time since they reach their optimal performance. This time is no less than four months, although significant results can be obtained before.

Actually, the best performance of a prediction program combines physical and statistical methods. As already said, physical models use statistics (MOS modules) to suppress systematic errors and to improve the model, and statistical models improve their results if a better wind input is used. In difficult terrain, to adapt the NWP to local conditions can lead to improvements of 10% in accuracy [22].

When it is necessary to forecast the production of the wind farms in a region, but there are no data available from all the wind farms present in it, it becomes necessary to upscale the results, i.e., to forecast the prediction for all wind farms, from the prediction for few of them. This process requires data from all the wind farms, in order to find the spatial correlation among all the wind farms present in the area. This is also a complex task, and a great amount of data is needed [11].

2.4. EVALUATION OF RESULTS

The error descriptors used for assessing the accuracy of the prediction are usual in Statistics. If we define the error in the power prediction for time $t+k$, produced at time t as $e(t+k/t)$, then this error has the value

$$e(t+k/t) = p(t+k) - \hat{p}(t+k/t)$$

Where $p(t+k)$ is the production of the wind farm at time $(t+k)$ and $\hat{p}(t+k/t)$ is the power predicted at time t for time $(t+k)$. Then, the error measures usually considered are:

- Normalized Root Mean Square Error ($NRMSE(k)$)

- $NRMSE(k) = \dfrac{1}{P_r}\sqrt{\dfrac{\sum_{t=1}^{N} e^2(t+k/t)}{N}}$
- Normalized Mean Average Error ($NMAE(k)$)
- $NMAE(k) = \dfrac{1}{P_r}\dfrac{\sum_{t=1}^{N}|e(t+k/t)|}{N}$
- Skill score ($R(k)$)
- $R(k) = \dfrac{\hat{\sigma}_P^2 - \hat{\sigma}_{e(t+k/t)}^2}{\hat{\sigma}_P^2}$

Where P_r is the rated power of the wind farm, N is the number of predictions examined along the considered time, produced k hours before the measure, $\hat{\sigma}_P^2$ is the estimated variance of the power production along the considered time, and $\hat{\sigma}_{e(t+k/t)}^2$ is the estimated variance of the error of the predictions produced k hours before the measure.

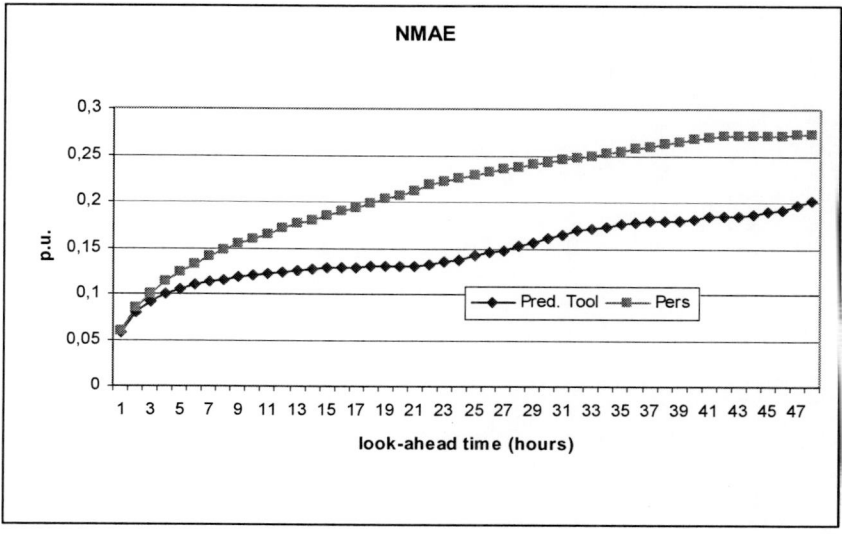

Figure 3. NMAE of a typical wind farm, for a prediction tool and persistence.

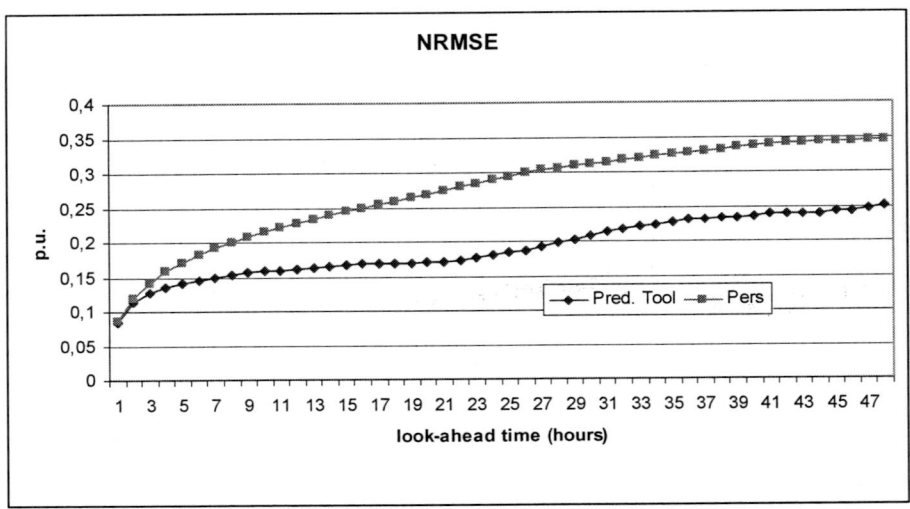

Figure 4. NMRSE of a typical wind farm, for a prediction tool and persistence.

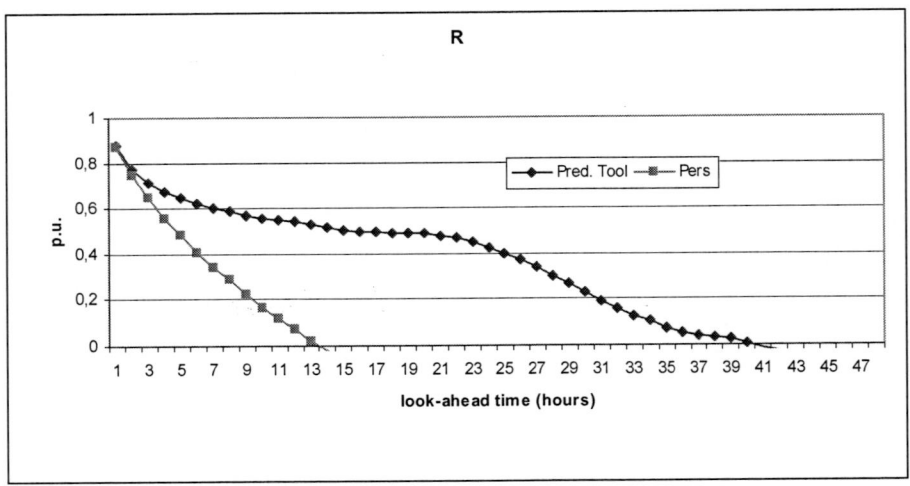

Figure 5. Values of R of a typical wind farm, for a prediction tool and persistence.

The skill coefficient tries to explain part of the variance of the data. It has a value between 0 and 1. A value of 1 means perfect prediction. A value of 0 means useless prediction – the model does not give any improvement compared with the constant mean model. It may be seen that all these error measures depend on the time elapsed between the moment when the prediction is generated and when the power is measured. This error increases when this time is bigger.

Values of *NMAE*, *NMRSE* and *R* for a wind farm of 21 MW are given in Figure 3, Figure 4 and Figure 5, for a prediction tool and persistence. It can be observed that the method always performs better than persistance, although for the first hours this difference is very small.

Some other features of wind power predictions, is that they are often biased, i.e., that the average value of the errors is not zero. Figure 6 shows this effect for eighteen months data of a wind farm of 14 MW of rated power. The error is defined as above. The mean value is always negative, so the bias is towards overprediction. This bias increases quickly the first hours, and then keeps stable.

Standard deviation also increases with look-ahead time, because these predictions are more inaccurate. The trend of standard deviation for the same wind farm is shown in Figure 7.

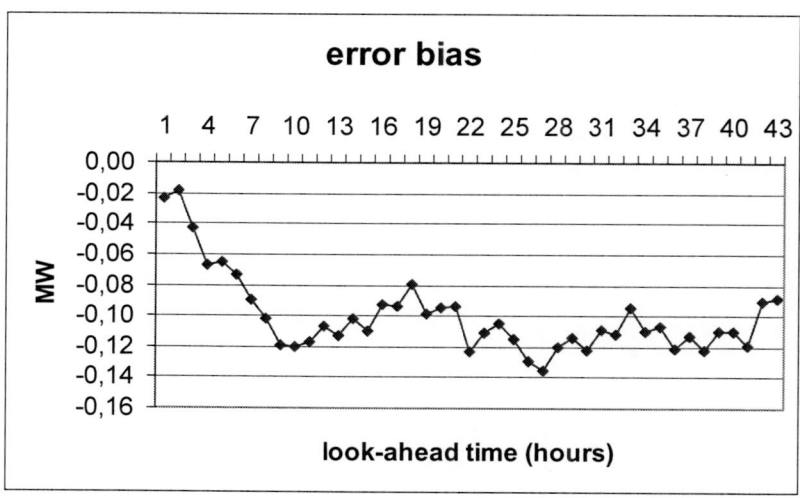

Figure 6. Error bias of short term wind power prediction.

The distribution of error has also skewness, as shown in Figure 8. This means that there are larger errors that are greater than the average.

All these results may be considered as typical, although the accuracy of wind power prediction may vary widely between different wind farms. This accuracy depends heavily on the roughness on the terrain, which makes wind predictions more difficult and inaccurate.

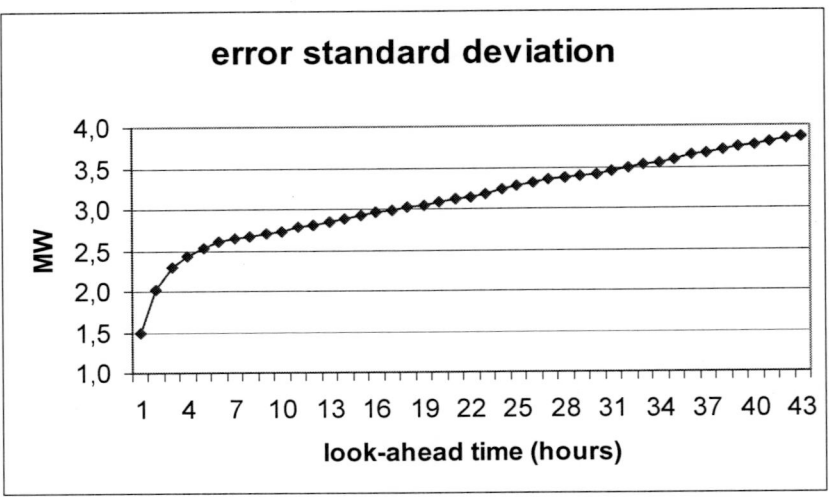

Figure 7. Standard deviation of short term wind power prediction.

It is also a common feature that the error decreases for a group of wind farms. Errors tend to compensate and give a better joint result. This is usually called *portfolio effect* that may increase the accuracy up to a 100%.

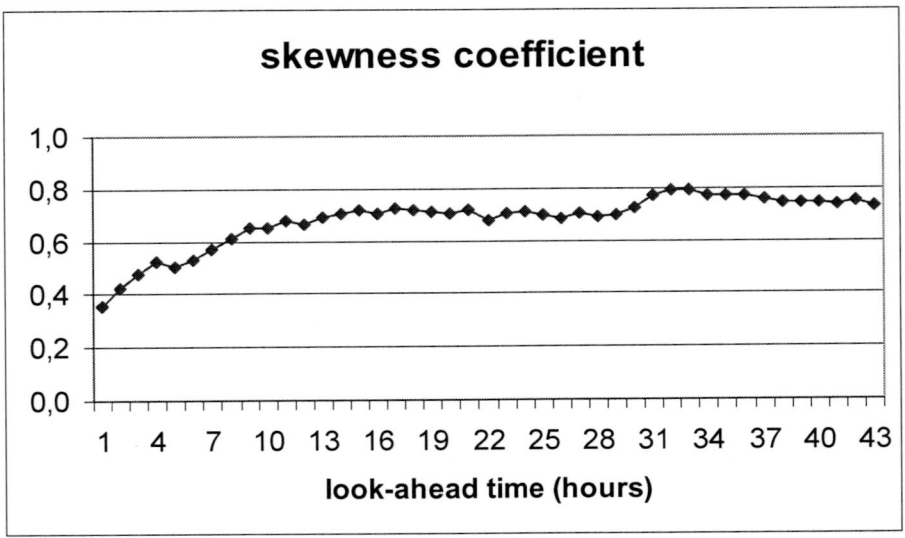

Figure 8. Skewness coefficient of short term wind power prediction.

2.5. SIPREÓLICO AS AN EXAMPLE OF A SHORT TERM WIND POWER PREDICTION PROGRAM

In this section, a more detailed description of a short term wind power prediction program is given. This intends to give a deeper insight into the properties, the accuracy and the application of these tools as a help to power system operation.

2.5.1. Introduction

Wind power has grown very quickly in the last years in Spain. The installed power in the Spanish peninsular system reached the value of 11854 MW in May 2007, and the growth of the power installed in the whole of Spain has been very quick, as shown in Figure 9.

Due to this amount of installed power, electric system operation must take into account the wind power production for the day ahead scheduling (grid constraints solution) and for real-time operation. This is why it has been revealed necessary a short term wind power prediction for the wind production in the whole of Spain. The Spanish System Operator, Red Eléctrica de España uses since the year 2002 the short term wind power program SIPREÓLICO, developed by Universidad Carlos III de Madrid [29], [30], [36], whose main features are described below.

SIPREÓLICO issues predictions each 15 minutes for the whole Spanish peninsular system. This means that predictions are generated for 88 farms or groups of farms, and then grouped by zones. Finally, the prediction for the whole peninsular system (more than 11.5 GW) is found from the individual predictions.

SIPREÓLICO receives numerical weather predictions four times a day from the National Meteorological Service of Spain (Instituto Nacional de Meteorología), who produces them with the program HIRLAM (see below). Some features of these predictions are given in Table 1.

As SIPREÓLICO has to manage a great number of wind farms, no detailed description for each of them, and consequently, no downscaling is possible by the moment, due to the huge computation times that would be needed. However, the portfolio effect keeps the error for the whole peninsular Spain within adequate levels.

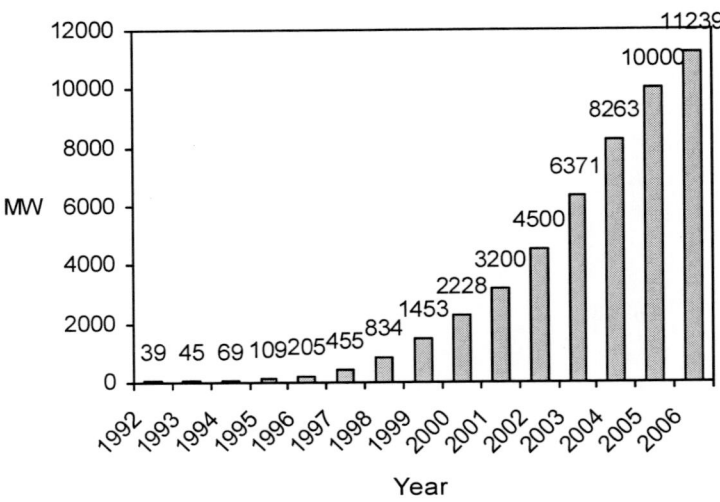

Figure 9. Installed wind power in Spain.

Table 1. Numerical weather predictions supplied by HIRLAM

Time of day (hour)	00	06	12	18
Time horizon (hours)	48	24	48	24
Spatial resolution (°)	0.16	0.05	0.16	0.05

Data of wind farms include information about the wind farm situation and rated power. There are not details about each wind farm. Wind farms are not obliged by the moment to provide the real time production to the System Operator. Hence, real time SCADA data used as input to SIPREÓLICO come from the distribution lines that evacuate only wind power. This is more than 80% of the total installed wind power. Due to this fact, it is not necessary any complex upscaling models. This upscaling just follows a proportional rule.

2.5.2. Prediction Algorithm

The core of SIPREÓLICO is the time series module. The design of it is critical, since it has to deal with changing relations between wind and power, which are nonlinear, and with numerical weather predictions that may have gross errors due mainly to the low spatial resolution of the NWP. This is why SIPREÓLICO uses different statistical models working in parallel, and only a combination of the most accurate among them are selected for each farm and each hour. The models used are the following:

- Univariant models. Autorregresive models AR(3), that take into account the wind inertia, and the daily cycle which is usually present in wind speed. These models can be used up to 10 hours before real time.
- Parametrical models with dependence of wind and speed direction. These models include the effects mentioned above in the univariant models, as well as the numerical predictions.
- Non parametrical models dependent on wind speed and direction. Nonparametric methods try to find the relations between the input to the model (the numerical weather predictions) and the output (the power produced by the wind farm). These relations depend on the time lag between the prediction and production, the wind speed and direction and the time in the day (diurnal cycle). These relations are found from past results and data, and are locally linearized. Since past predictions and outputs do not define clearly a curve, but values around each point, a probability distribution function is obtained, from which the prediction is finally issued. An illustration of this model is given in Figure 10.

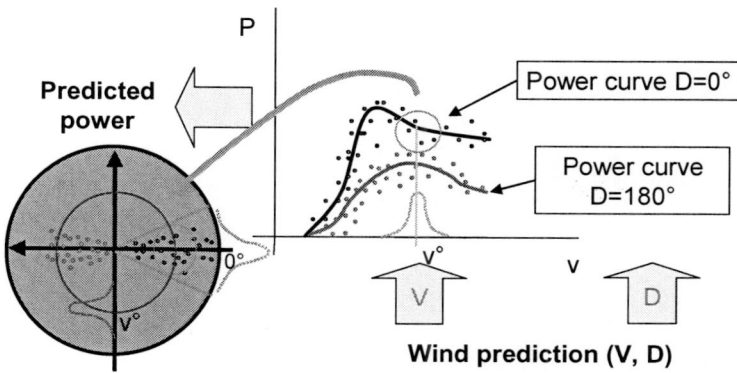

Figure 10. Nonparametric methods.

However, for an accurate prediction, two more aspects must be considered: adaptative estimation of the time series parameters, and the optimal combination and selection of the best prediction provided by the different tools.

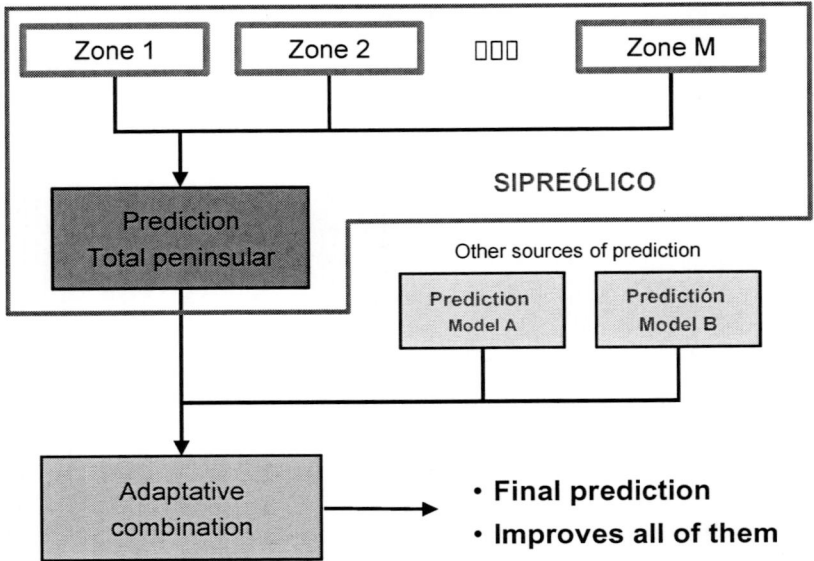

Figure 11. Scheme of SIPREÓLICO operation.

The parameters of the models change with time, due to seasonal variations of weather, wind farm repowering, or other reasons. Hence, these parameters must be adapted with the new data that are being obtained. This parameter updating is made using two alternative procedures, Recursive Least Squares and Kalman Filter.

The final prediction is made by a final combination of the parallel predictors. This combination is based on the actual performance of each predictor. For this reason, it is necessary to implement a real time evaluation of each of them.

Then, the degree of performance of each predictor in each horizon determines the optimal weighting of each forecast in the final predicted output, together with several measures and diagnostics. The evaluation of the predictions of each predictor is made with an exponentially weighted mean squared prediction error (EW-MSPE). The final prediction for a given horizon is obtained using the combination of the 3 predictors with lower EW-MSPE. A more detailed description may be found in [30].

Finally, this prediction is optimally combined to prediction coming from other sources, and the accuracy of the joint prediction is higher than any of the incoming sources. This process is illustrated in Figure 11.

2.5.3. Application in Red Eléctrica de España

The wind generation forecasts produced by SIPREÓLICO are already used in the following aspects of the system operation:

- Improvement of the day-ahead demand prediction carried out by the Control Centre of Red Eléctrica de España.
- Determination of the reserve needs for the secondary and tertiary regulation markets.
- Continuous resolution of the technical constraints of the system.
- Transmission grid works and line disconnections scheduling

When a line disconnection takes place in a zone with many wind farms, problems may arise for the evacuation of the whole power generated in the area (this could be the case when both the generation groups and the wind farms are working at full power). As an example, it may be said that a situation such as this took place in the Northwest of Spain during the disconnection of a 220 kV line. Since wind generation forecasts were available 48 hours in advance, the system operator imposed a limitation to the technical minimum to several conventional groups in the area, when necessary. A study conducted later showed that the decisions adopted/undertaken by the system operator thanks to SIPREÓLICO forecasts were corrected in a 96% of the cases. More detailed information may be found in [14].

2.5.4. Accuracy of Results

In this section, the accuracy of SIPREÓLICO is shown. The evaluation of this accuracy for the last month before June 5, 2007 are shown in the following figures, that corresponds to the whole of the wind power connected to the Spanish peninsular grid. These results have been kindly provided by Red Eléctrica de España. The high accuracy of these results is partly due to the portfolio effect applied to all the generation connected at that moment.

2.6. OTHER PREDICTION PROGRAMS

Many other short term wind power prediction programs are currently being used, and are available. In this section a few of them are shortly described.

WPPT [23] (Wind Power Prediction Tool) is a prediction program developed in Institut of Mathematical modelling of the Technical University of Denmark and it is running since 1996 for different users. It is a statistical model, and uses non-linear time series models, including a semi-parametric power curve model for wind farms taking into account both wind speed and direction. The models are self calibrating and adaptive. The inputs are predictions of wind speed and direction as well as real time power production measurements and other meteorological variables. It produces predictions for up to 120 hours ahead, depending on the input predictions.

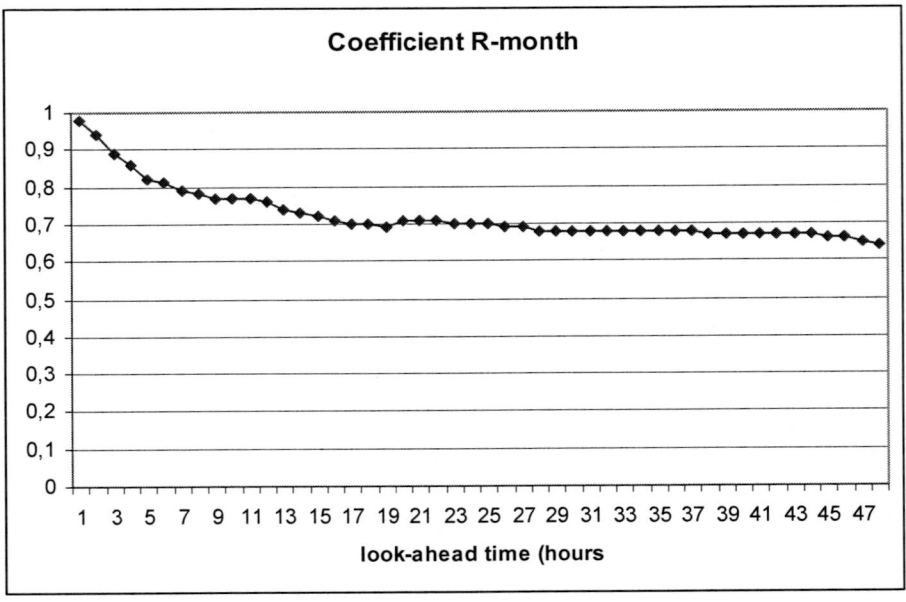

Figure 12. Value of coefficient R for the last month in the Spanish peninsular grid.

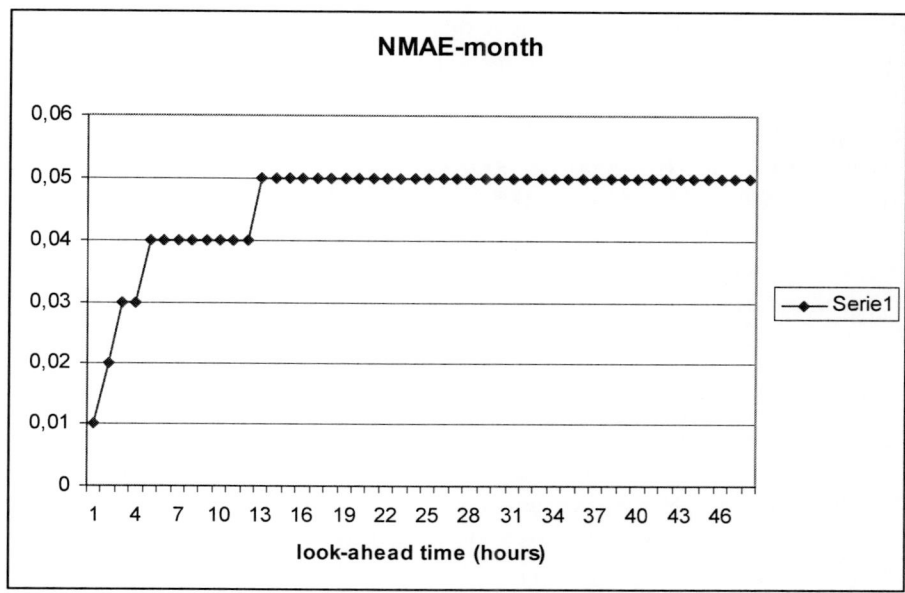

Figure 13. Value of NMAE for the last month in the Spanish peninsular grid.

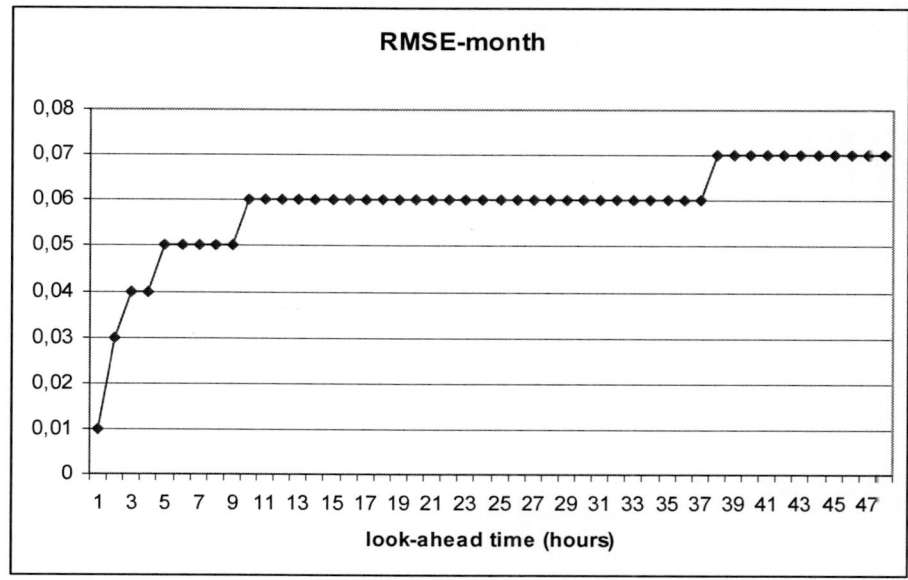

Figure 14. Value of RMSE for the last month in the Spanish peninsular grid.

Prediktor [19] is a prediction tool that uses physical models and MOS. Its basic input are the NWP, and information about the wind farm. The NWP are adapted to local conditions, taking into account the turbines siting, the orography, the roughness of the terrain, etc. MOS (model output statistics) are used to correct the model. The observed wind farm productions are used to adjust the parameters of statistical filters. The final output of the model is the expected production of the wind farm every 3 hours over the next 48 hours. It has been developed by Risoe, and has been used in Denmark, the USA and elsewhere.

LocalPred [21] is another tool that uses physical model that includes weather prediction using mesoscale models. Its inputs are the NWP and data for the wind farm, such as orography, roughness, turbine siting, and past data. MOS statistics and real production are also used for tuning the model. It has been developed by CENER and CIEMAT and is currently being used in Spain.

Advanced Wind Power Prediction Tool (AWPT) [10] has been developed by ISET, Germany. It uses statistical models using Artificial Neural Networks. They use NWP and a few measurements from selected wind farms for producing predictions of a large area in Germany with more than 13 GW of wind connected. Off-line measurements of the production in the region are used to tune the upscaling method.

eWind [3] is an USA product, developed by AWS Truewind that produces predictions using physical modelling. From NWP produced by a Regional Weather Model, mesoscale simulation tools adapt the prediction to local conditions. MOS are also used to tune the models and correct systematic errors. This model is being widely used by wind farm owners in the USA.

Previento [11] is another physical prediction model developed in the University of Oldenburg that takes the prediction inputs from different services. The prediction of regional production is upscaled from selected wind farms, and MOS are used to correct the model results. It is being used in Germany.

ANEMOS is the result of an European project that joins different prediction tools in Europe. The project aims to develop accurate models or improve existing ones, aiming to outperform actual state-of-the-art, for onshore and offshore wind resource forecasting (statistical and physical). Emphasis is given on integrating high-resolution meteorological forecasts. For the offshore case, marine meteorology is considered, as well as information by satellite-radar images. ANEMOS itself is an integrated software that hosts the various models developed by the project's partners. The system is being used by several utilities and wind farm owners for on-line operation at onshore and offshore wind farms for local/regional/national wind prediction. The applications are suited to different terrains and climates, on-/near-/off-shore farms, interconnected or island grids. A detailed description of this project may be found in [16].

More information about the state of the art in short term wind power prediction may be found in [12]. An interesting study (in Spanish) comparing results for different wind farms and prediction tools in Spain is [2].

The current trend of research in the prediction field is focused on a better use of mesoscale models, the combination of different predictions coming from independent sources, and the use of ensembles [13]. These lines are very promising, and future increases in accuracy can be reasonably expected. Besides, the advances in computation will also lead to NWP more frequent and with greater resolution and accuracy.

Chapter 3

BIDDING STRATEGIES FOR WIND FARM OWNERS

One of the applications of short term wind power prediction programs is to help the wind generators to participate in the electricity markets. As shown in the next sections, to use such programs implies important benefits to the users. Participation of wind energy in electricity market is now becoming usual, and this is why this section aims to quantify those benefits and describes the possibilities of using these tools, pointing at the consequences of this use.

3.1. WIND POWER AND ELECTRICITY MARKETS

Electricity markets are nowadays the most common paradigm for electricity trading. In many of the electricity markets, a important amount of the energy is traded in a daily market, that may account for the 20 to the 80% of the total energy, and the rest is traded bilaterally between generators and consumers. To participate in these daily and intraday markets, it is necessary to present bids for selling or buying power at a certain price, more than 12 hours before the operation time. The energy is committed to each market participant by means of an auction. To deviate from this commitment may suppose an overcost for the system, and must be paid accordingly. However, it is possible in some countries to trade the foreseen power excess or deficit in intraday markets run along the day, if this imbalance is known several hours beforehand. Other countries, such as the UK, have a different market model, with a small balancing market that is closed shortly before the operation time [25]. Table 2 summarizes some of these features for different European countries.

Wind energy must integrate into electricity markets, in spite of its non-dispatchable nature, and the variability of the resource. Also, wind energy, as coming from a renewable source of energy, usually benefits from subsidies. These subsidies may take the form of feed-in tariffs, green certificates of other mechanisms, such as tenders, that vary widely among the different countries.

It is not common that wind energy participates in electricity markets. Wind generators are usually allowed to inject all their power to the grid under preferential rules, and they are just paid the feed-in tariff for their production. In order to assess the wind power production, TSOs and balancing companies must run prediction programs for a region in order to allocate reserves, to account for the possible congestions in the system, to buy the energy need by their customers, etc.

However, there is a trend to encourage the participation of wind power in the electricity markets, following the market rules as any other participant. This means that wind generators must submit bids to the Power Exchange office, the day before operation, saying how much power they want to sell, and its price. Since the power generated cannot be easily stored, the price is usually zero, and all the bids are accepted. However, the amount of power committed in the energy auction must be generated, or the system overcost due to deviations from this commitment should be paid. For this reason, wind power generators participating in an electricity market need to make power predictions between (typically) 12 and 36 hours in advance of the operation time for bidding in the daily markets. Whenever possible, it is advisable to negotiate in intraday markets using updated predictions, in order to reduce the imbalance cost.

Participation of wind generators in electricity markets has several advantages. Predictions made by wind generators themselves may be more accurate than those produced at a regional scale, since they have more information about their wind farms. For instance, they know accurately the orography of the terrain, and a downscaling may improve the NWP input, and they know better the availability of the wind turbines for the next day. This information arrives with more difficulties to a regional organization. In general, a better prediction helps the system operation.

Table 2. Common points in regulatory frameworks

	Pool	Bilateral contracts	Gate closure	Intraday/balance markets
Spain	X		14 h	Yes (2h15m)
Denmark	X	X	12 h	Yes (30 m)
France	X	X	13 h	Yes (2 h)
Ireland	X	X	14 h	Yes
Eng.&W		X		Yes (1 h)
Greece	X	X	12 h	No

3.2. BENEFITS OF USING SHORT TERM WIND POWER PREDICTION TOOLS

The wind generators that participate in an electricity market obtain interesting benefits from using a short term wind power prediction tool, because the amount of power imbalance reduces significantly. In this section, this reduction is quantified under rather general assumptions.

In general terms, the revenue R for a given wind farm in a pool market may be generalized as:

$$R = \sum_t P_{d,t} \cdot MP_t + \sum_t MP_{i,t}(P_{i,t} - P_{di,t}) + imbalance \tag{1}$$

Where

$$imbalance = \begin{cases} +MP_t^{up}(P_{gen,t} - P_{last,t}) & P_{gen,t} > P_{last,t} \\ -MP_t^{down}(P_{last,t} - P_{gen,t}) & P_{gen,t} < P_{last,t} \end{cases} \tag{2}$$

The meaning of the different terms of the equations is:

$P_{gen,t}$ Power actually generated by the wind farm in the hour t
$P_{d,t}$ Power committed to the wind farm in the daily market for the hour t. It coincides with the prediction available at the gate closure of the daily market.

$P_{i,t}$ Power committed to the wind farm in the intraday market for the hour t It coincides with the prediction available at the gate closure of the intraday market.

$P_{last,t}$ Power committed to the wind farm in the last update for the hour t

MP_t Marginal price of energy in the daily market for the hour t

$MP_{i,t}$ Marginal price of energy in the intraday market for the hour t

MP_t^{up} Marginal price of energy in the spot market for selling energy in the hour t

MP_t^{down} Marginal price of energy in the spot market for buying energy in the hour t

In markets without intraday markets, the second term of equation (1) does not apply. In these cases $P_{last,t} = P_{dt}$. Where there is a possibility for updating bids in an intraday market, $P_{last,t} = P_{it}$. An example of time schedules for daily and intraday markets is shown in Figure 16.

The imbalance term (2) is simpler for some markets, because the price in both senses of the imbalance is the same. In other markets, this price depends on the system deviation and may be zero in some cases. Most times the imbalance term represents a decrease in revenue, and an accurate prediction program should reduce as much as possible this term.

In order to simplify the equations, for a better understanding of the results, some simplifications are going to be made to them.

Equations (1) and (2) are simplified to:

$$R = \sum_t P_{gen,t} \cdot MP_t - \sum_t P_{dev,t} \cdot \psi \cdot MP_t$$

where

$P_{dev,t} = |P_{gen,t} - P_{d,t}|$
ψ penalty factor.

Hence, the following simplifications have been assumed:

- The intraday market has not been considered. Since the overpredictions and the underpredictions could be assumed to be more or less unbiased, and the intraday MP might be equally higher or lower than the daily MP,

the weight of this term is not high. The assumption of unbiased predictions seems not to be completely correct as shown later.
- The imbalance term has been changed into a penalty proportional to the absolute value of the deviations incurred, and the amount of the penalty is a fraction of the marginal price. This simplification has the following effects in different markets:

For those markets where the penalties are paid only if the imbalances have the same sense than the overall system deviations, this assumption is a pessimistic one. Better results must be expected.

For those markets with different penalties (or market prices) for up deviations and down deviations, the effect is small, due only to the bias of the prediction.

The results presented in this section compare the performed prediction and persistence with perfect prediction. The comparisons are made between the revenues of a producer that would perform a perfect prediction and those of a producer using SIPREÓLICO as a prediction tool. This comparison has been made also for predictions performed using persistence. The value of the relative revenue (RR) that compares these values is given by the following equation.

$$RR = \frac{\sum_t P_{gen,t} \cdot MP_t - \sum_t P_{dev,t} \cdot \psi \cdot MP_t}{\sum_t P_{gen,t} \cdot MP_t} \qquad (3)$$

The lower part of the equation represents the revenues obtained with perfect prediction. The upper part is the loss in revenues due to inaccuracies of the prediction tool. Therefore, the term RR will be always less or equal than 1.

A further simplification is to try to make equation (3) independent of the price. If we consider that it leads to the same results for a long period of time to be paid every hour the MP or to be paid the average MP along that period, then the term RR of equation (3) changes into $RRapp$ of equation (4).

$$RRapp = \frac{\sum_t P_{gen,t} - \sum_t P_{dev,t} \cdot \psi}{\sum_t P_{gen,t}} \qquad (4)$$

In this equation it may be seen that the reduction income depends proportionally of the penalty ψ. Results have been obtained for different values of this parameter.

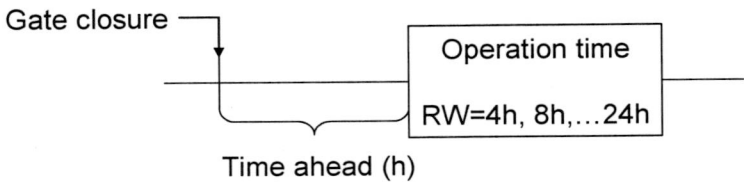

Figure 15.

The study has been performed for different rolling windows (RW), and for different times between gate closures and operation time (Ta). An illustration of these times is given in Figure 15.

Some additional assumptions concerning market rules are made:

- A pool system is considered. Results under bilateral contracts would depend on agreed conditions, difficult to know. However, bilateral prices and spot prices could be assumed as correlated in the long term, so that the conclusions could be extrapolated for markets based on bilateral contracts.
- The use of past market prices implies to assume that wind power has no influence on market prices. This is a too simplifying assumption, especially for countries with a high wind power penetration. However, it must be considered that most farm owners are price takers in a competitive market, and for them the prices are independent variables. The conclusions are intended to be independent of the price level.
- There is no difference between wind generators and conventional generators in the market. That is to say that no subsidies are considered for the wind generation. The losses due to imbalance costs then are smaller in relative terms than those shown here.
- Wind generators present their bids for the amount given by the prediction program, at zero price, so they are always accepted.
- The prediction tool makes new prediction from available data (wind forecasting and real-time production) every hour.

- SIPREÓLICO has been the prediction tool used for performing the prediction. This prediction tool may be considered as representative of the state of the art.

These equations have been applied to offline data of two different real wind farms, with the following features:

WF 1: 19 months of power measurements, farm situated on flat terrain
WF 2: 3 months power measurements, farm placed on mildly rough terrain

The Spanish market prices have been used for the studies. However, the results are fairly independent of the price level.

According to equation (4), the revenues' loss is an amount that depends on the accuracy of the prediction tool, and independent of the actual market prices. The average value for all considered values of time ahead and rolling window could be considered as a characteristic parameter of that farm that gives a measure of the advantages of a prediction tool at a glance. These results are given in Table 3.

Table 3. Losses as a percentage of the penalty factor

	WF1	WF2
SIPREÓLICO	39.37	23.52
Persistence	79.21	45.19

In this table, the number 39.37 means that for a penalty of 0.4 times the marginal price, the loss for WF1 would be 0.4x39.37=15.748% of the revenue, when compared to the revenues with prediction. As this number is the average for very different values of look ahead times and rolling windows, the losses may be smaller. More details of this study may be found in [34].

The effect of trading in the intraday market the difference between old predictions for the daily market and the new ones is also to reduce the cost of the imbalance. If, for instance, we consider that prediction for the daily market are produced at 1100 of the previous day, the look-ahead time for the predictions is between 13 and 16 hours. If there is an intraday market each 4 hours, with a market gate closure 3 hours before operation time, the updated predictions can be used with a look-ahead time between 3 and 6 hours, with a much better accuracy. This time schedule is shown in Figure 16.

Under these assumptions, the reduction of the imbalance costs when trading also in the intraday markets is shown in Figure 17.

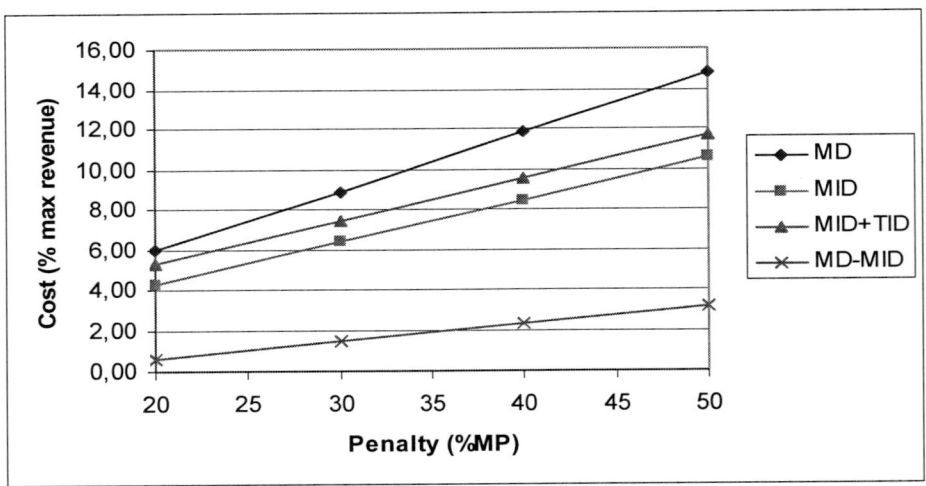

Figure 16. Time schedule of the different markets. D = daily market; ID = intraday market.

Figure 17. Effect of intraday markets on revenue. Results for WF2.

In this figure, the reduction in revenue due to imbalances is shown against the penalty, expressed as a percentage of the marginal price (MP). Different cases are considered – only bids in the daily market (MD), bids in the daily and updates in the intraday markets (MID), bids in the daily and updates in the intraday markets, plus the price of purchasing energy in the intraday market (MID+TID). The difference of the income reduction between MD and MID is also shown. The positive difference between (MID+TID) and (MID) comes from the bias that tends to overpredict rather than to underpredict, and then most of times energy must be bought in the intraday market.

It is apparent that the decrease in lost revenue may be quite high and obviously increases with time and frequency of intraday markets. For very low penalties, and if there is a bias for overpredictions in the prediction program, to participate in the intraday markets may imply only a small advantage, since the expenses in buying more energy are more or less the same amount than the reduction in penalty for imbalance.

3.3. UNCERTAINTY OF WIND POWER PREDICTIONS

The results shown in the previous sections have been obtained by considering only one value produced by the prediction program used.

However, the predictions provided by a short term wind power prediction program are uncertain, and it is interesting to estimate this uncertainty to have more information about the future production of a wind farm. As we will see later, this knowledge may have economic consequences for the wind farm owner.

Let p be the random variable associated with the power output of a wind farm. Then, the probability of producing p MW, having predicted p^* MW k hours before, is given by the probability density function $f_{p^*,k}(p)$.

The uncertainty, and hence the probability density function, changes with the range of the wind farm power output, since this value is bounded between zero and the rated power. Besides, the power curve of a wind turbine or wind farm is nonlinear. If we assume that the wind power predictions have gaussian uncertainty, then the probability density functions of the power predictions will not be gaussian [18].

The shape of these probability density functions is also affected by the time lag elapsed between the prediction and the operation times. As shown before, predictions with a shorter time lag are more accurate, and their variance is smaller than those predictions produced longer before.

To obtain analytically, or in real time, the uncertainty of this prediction is difficult, but approximate estimations can be made from past data, and some research has already been made in this field [18]. Given the past predictions and wind production for these predictions, the accuracy of these predictions can be tabulated, and then their frequency can be used as an approximation of these probability density functions.

If the power range of a wind farm is comprised between 0 and P_{max}, and this range is divided in Q intervals, the power p is included in the interval q, if

$$\frac{q-1}{Q}P_{max} < p \le \frac{q}{Q}P_{max}$$

The probability density function $f_{p*,k}(p)$ may change into $f_{q*,k}(p)$, where q^* is the interval in which the predicted power p^* is included.

As an example, the following figures give the frequency distributions of the produced powers for different values and time lags of the prediction. Figure 18 reflects the frequency distribution when a low power had been predicted 7 hours before real time, while Figure 19 shows the frequency distribution when the power level is near the average.

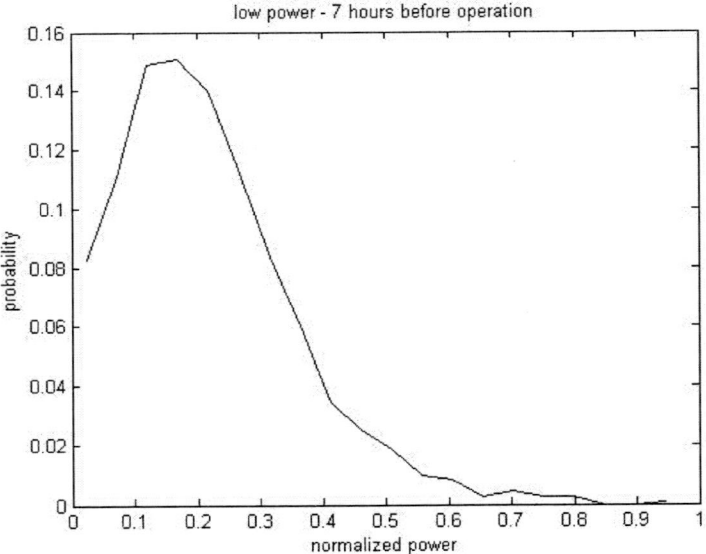

Figure 18. $f_{q*,k}(p)$ for $q^* = 2$ and $k = 7$. $Q = 14$.

Figure 20, on the other hand, shows the frequency distribution for a prediction in the high range of power 7 hours before the operation, while Figure 21 gives the frequency distribution for the same power range, but 36 hours before operation. It can be seen that this last distribution has a larger variance.

All these values have been obtained from real production of three months of a wind farm whose rated power has been normalized to 1.

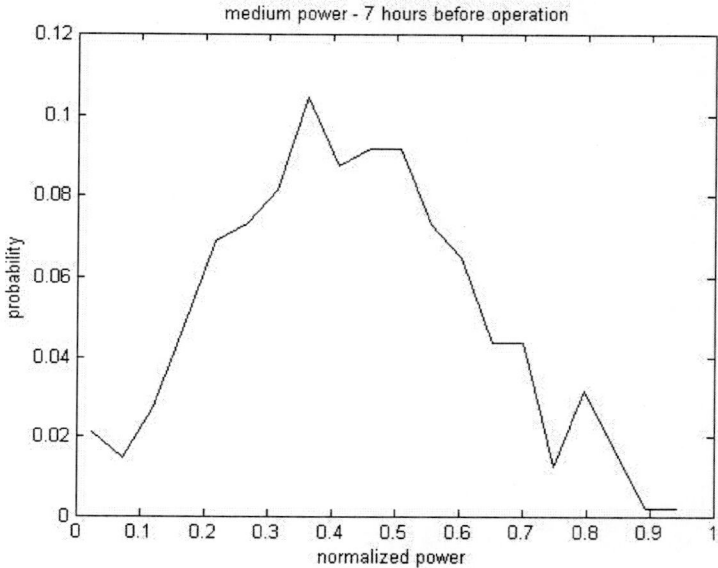

Figure 19. $f_{q^*,k}(p)$ for $q^* = 7$ and $k = 7$. $Q = 14$.

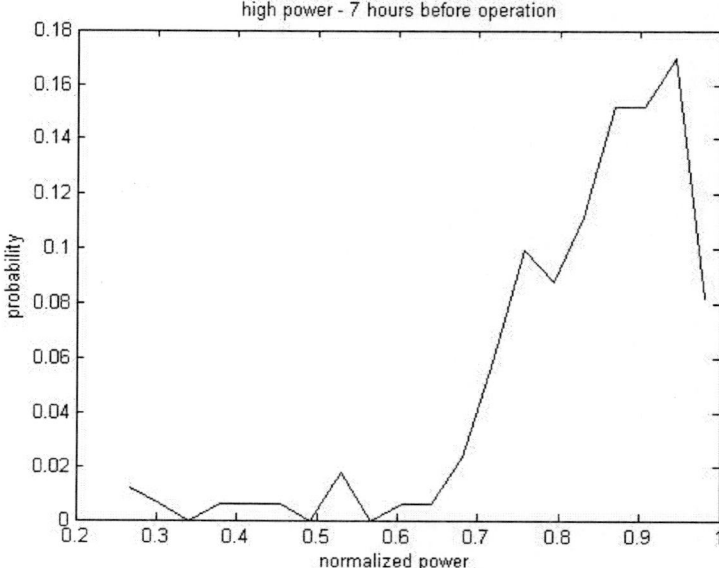

Figure 20. $f_{q^*,k}(p)$ for $q^* = 13$ and $k = 7$. $Q = 14$.

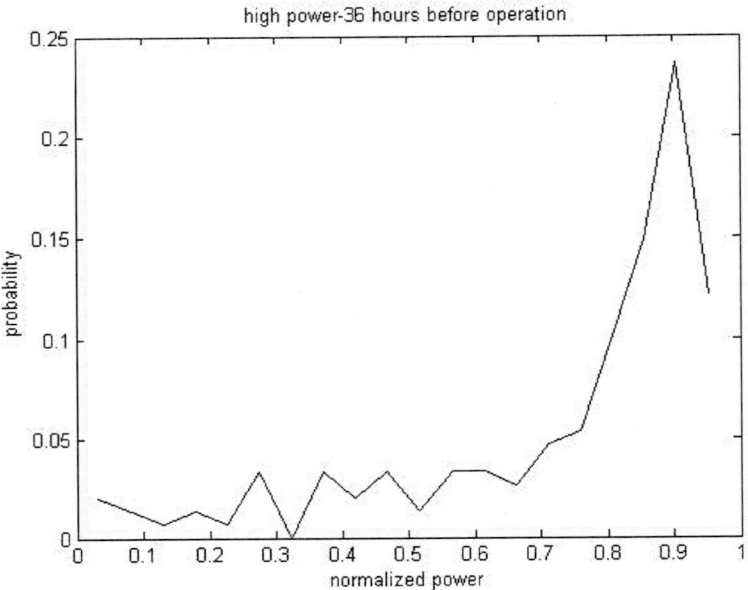

Figure 21. $f_{q^*,k}(p)$ for $q^* = 13$ and $k = 36$. $Q = 14$.

3.4. BIDDING WIND POWER UNDER UNCERTAINTY

The evaluation of the uncertainty of the predictions can be used advantageously in the bids' submission to the electricity markets. Due to the fact that the cost of imbalances are usually higher when the generator is "short" (it produced less than predicted, and energy must be bought) than when it is "long" (the production is greater than expected, and extra power is sold), to bid a different power than the average of the probability distribution (the usual output of a prediction program) may yield greater revenues. The following sections quantify this effect, that should be taken into account by System Operators and Regulators.

3.4.1. Problem Formulation

Since the predictions are more accurate when the prediction and the operation times are closer, a possible use of this uncertainty evaluation could be to bid the output of the prediction program (average of the distribution) in the daily market (13 hours before operation), and to correct these bids in the intraday, according to the expected price of the energy in this markets, the expected penalty, and the uncertainty of the more recent predictions. This is developed in this section.

As explained in a previous section, the revenues of the wind farm may be calculated by the formulas (1) and (2). When bidding in the intraday markets, the bid to the daily market (the committed power) is already known, as well as the price for the energy in this market. The price of the intraday markets is unknown, although they can be predicted, as well as the price of the imbalance. At the time of the intraday market, a prediction is generated for the next hours, and this prediction has an uncertainty estimated from past data as explained in last section. Then, the revenue for a given time could be expressed as a function of the power bid (or power traded) in the intraday markets, and the power actually generated, as $R(P_{last}, P_{gen})$, where P_{last}, and P_{gen} are vectors of bid and generated powers for a whole period (one day, or a few hours).

The aim of the problem is to obtain the value of P_{last} that maximizes the revenue for a given set of intraday energy prices, imbalance costs and the uncertainty of the prediction. This problem may be formulated for each hour as in eq. (5). The formulation for more hours is straightforward. Since time couplings between the different hours are not considered, this formulation is general.

$$\max_{P_{last}} \sum_{j=1}^{j=Q} \left\{ R(P_{last}, P_{gen,j}) \cdot f_{q^*,k}(P_{gen,j}) \right\} \tag{5}$$

In equation (3), Q is the number of intervals of the power range of the wind farm that have been considered in the uncertainty probability density function, and $f_{q^*,k}(P_{gen,j})$ is the probability of the generated power to take the value $P_{gen,j}$ when the prediction k hours before is in the interval q^*.

The assumed hypotheses for the study have been the following:

- A pool system has been considered. Wind producers make their bids for a given amount of power at price zero. This means that bids are always accepted.
- The prediction of the prices of the intraday market is perfect.
- The subsidies for wind energy are not considered.
- The prediction tool makes new prediction from available data (wind forecasts and real-time production) every hour.

- SIPREÓLICO has been the prediction tool used for performing the prediction.

3.4.2. Study Case

In order to show the consequences of taking into account the uncertainty of the prediction into the power bidding, an example has been run with real data from wind farms and electricity prices.

The data of wind farm come from the actual production of a wind farm of 14 MW of rated power during three months. The probability functions of the wind farm have been obtained from these same production data and predictions performed for this wind farm for these three months.

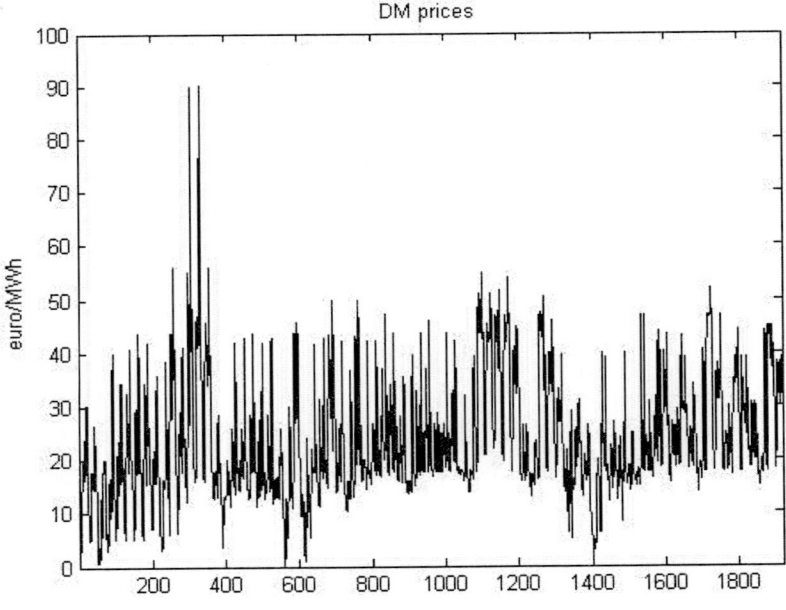

Figure 22. Prices for the period considered.

Although the study conditions do not follow the Spanish market rules, the prices of the Spanish market between January and March 2003 have been used for this study. The level of the prices is given in Figure 22. The average price for this period was 23.678 c€/MWh. The intraday market prices for this period where also used. The average intraday market price used was 22.4791 c€/MWh.

The study performed includes a comparison between three different assumptions:

1. OPTIMAL: The described method of maximizing the revenues taking into account the uncertainties of the wind power prediction tool.
2. BEST PREDICTION: When the best prediction available is used to modify the bid in the intraday markets.
3. NO INTRADAY: When no updating is produced in the intraday market. Therefore, only one bid per day is produced.

The study has been performed for eighty days. Longer studies, however, do not lead to very different results.

3.4.3. Results

For the three assumptions, and for values of $MP^{sell} = 0{,}5 \cdot MP$ and $MP^{buy} = 1{,}5 \cdot MP$, the revenues, the average errors and the average of the absolute value of the errors along the whole period are presented in Table 4. It can be seen there that the most profitable option presents larger errors than the most accurate. Errors and absolute errors for instant t are defined as:

$$error_t = P_{last,t} - P_{gen,t}$$
$$abs\_err_t = \left| P_{last,t} - P_{gen,t} \right| \tag{6}$$

Table 4. Errors for the different assumptions

	Optimal	Best prediction	No intraday
Revenues (€)	290 112	280 500	269 046
Error (MW)	-0.3661	0.1412	0.1619
Absolute error (MW)	1.8722	1.4132	2.0198

Although the values of MP^{sell} and MP^{buy} are realistic in the sense that, usually, to buy energy at the last moment is more expensive than to sell it, real systems have more complex methods to calculate the cost of the imbalances.

Figure 23 shows the differences between the three assumptions. The asterisks (*) show the difference between the OPTIMAL and the BEST PREDICTION assumptions. The diamonds show the difference between the OPTIMAL and the NO INTRADAY assumptions. The dotted line shows the difference between the BEST PREDICTION and the NO INTRADAY assumptions. It may be observed that the differences, especially between OPTIMAL and BEST PREDICTION are almost always greater than zero. Only in some cases where the predictions had been bad, the results for the BEST PREDICTION were better than the OPTIMAL assumptions.

However, if we check the error between the actual generated power and the different powers, as shown in Table 4, we can see that, even if the errors of the OPTIMAL assumptions are greater, the revenues obtained are higher, and then, the bids provided by the wind farm owners aiming to maximize their revenues are not the most accurate. In Figure 24 a sample of the power error in a series of 30 hours is shown. It may be seen there, that the error of the best prediction is lower than the optimal, and also lower than the case where only bids to the daily markets are presented. Although there are cases where this does not happen, the general behaviour follows this pattern.

Table 5. Revenues for different sell and buy prices (€)

Buy/Sell prices	1.5/0.5	1/1	0.5/1.5
Optimal	290 112	348 834	467 178
Best prediction	280 500	311 751	343 002
No intraday	269 046	313 039	357 032

From Table 4 we can deduce that the prediction tool has a trend for overprediction as expected, because the errors, defined as in (4) are positive. This is more apparent when the buy and sell prices change from those assumed in the previous results.

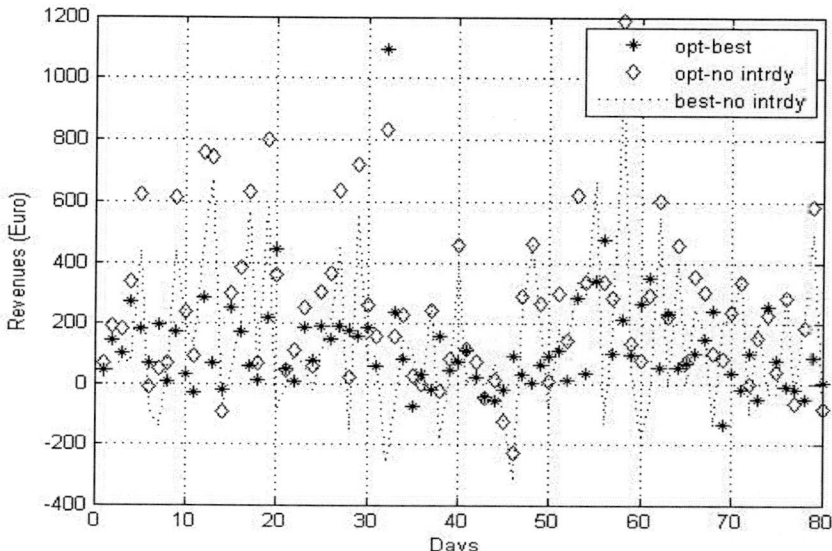

Figure 23. Results for one month of the cases studied.

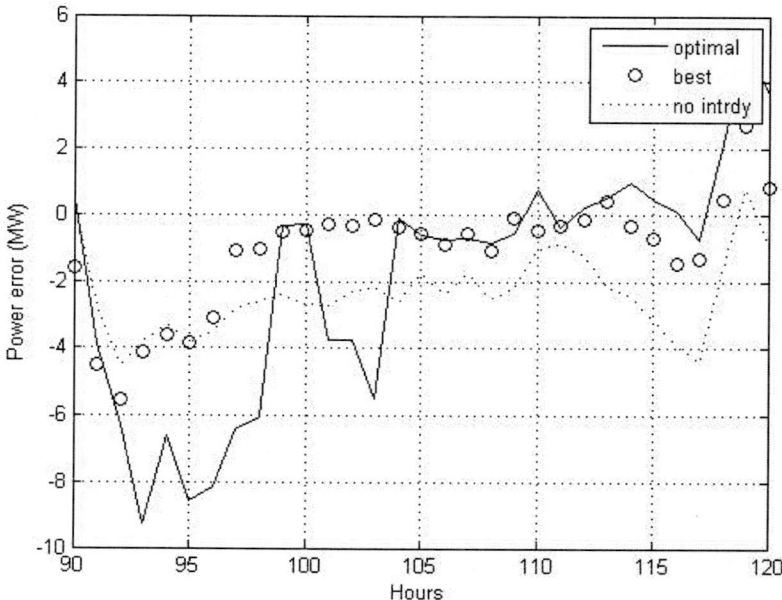

Figure 24. Power errors for a series of 30 hours.

Although the following discussion is mainly of theoretical interest, it gives a good insight into the properties of the process. The Table 5 gives different results for different values of sell and buy prices. The second column has the same values as the first row of Table 4, but it is included here for an easier comparison.

We can see in this table that:

1. The revenue when bidding the optimal power is always the highest of the three possible bids.
2. The revenue is higher when the sell price becomes greater. This means that the system tends to overpredict, and then, power must be bought at the last moment most of the times. This need of purchasing is an additional cost that reduces the overall revenues.
3. When the sell price is equal or lower than the buy price, the revenues are larger when no updates are made in the intraday markets. This is also a consequence of the tendency to overpredict, since the error is higher in this last case, as shown in Table 4.

Table 6. Average errors for different sell and buy prices (MW)

Buy/Sell prices	1.5/0.5	1/1	0.5/1.5
Optimal	-0.3661	-0.2423	0.6255
Best prediction	0.1412		
No intraday	0.1619		

Table 7. Average absolute errors for different sell and buy prices (MW)

Buy/Sell prices	1.5/0.5	1/1	0.5/1.5
Optimal	1.8722	4.2822	5.5945
Best prediction	1.4132		
No intraday	2.0198		

In order to show the bias of the errors and absolute errors, these errors are given in Table 6 and Table 7. In them, the values for the BEST PREDICTION and NO INTRADAY assumptions given in Table 4 are also included for an easier comparison.

From these tables we may conclude also that:

- The lowest errors are always given by the BEST PREDICTION case. This better performance, however, does not lead to the highest revenues, even compared with the NO INTRADAY option, if the sell price were higher than the buy price.

- The errors from the OPTIMAL case show that with this strategy, the trend to overpredict of the prediction tool is compensated. When the sell price is higher than the buy price, the bid tends to be higher than the actual generation. The fact that the average error is negative when both prices are equal is another consequence of this trend to overprediction.

Chapter 4

OPTIMAL DISTRIBUTED MANAGEMENT OF WIND POWER

Geographic distribution of wind farms depend mainly on best wind resource locations and environmental regulations. Therefore, wind parks are generally concentrated in zones of the country, where these installations are more profitable. In addition, wind farms should be connected to the transmission system in buses which can accept non-dispatchable generation in a secure way. For this, it is frequent to find two or more wind farms injecting power to the same transmission system node. The wind farms are connected to the transmission bus through a subtransmission grid, generally including only wind farms. This is the case in Spain, where the larger installed wind power (4922 MW, 42.4% of the full wind power installed capacity) is allocated in the North-West side of the country (2007 data, [38]). In this region, up to 6 wind farms of an average power of 30 MW may inject active power to the same transmission node. The geographic closeness among the wind producers and their particular production characteristics have led to create regional Renewable Control Centres or Delegated Dispatches (DD's), to control the wind farm production. At the moment, DD's only can include wind power generators linked to the same transmission bus and to proceed in contingence situations. Spanish System Operator (REE, *Red Eléctrica de España*) had created a centre called CECRE (*Centro de Control para Régimen Especial*), to manage the renewable production in contingency situations. CECRE coordinates the action of DD's and independent wind farms (not included in any DD) in contingency situations, receiving information about the production of the wind farms and setting the maximum admissible active power injections of wind power in the nodes of the transmission system.

In the next sections, an optimal approach for the DD control actions in contingency situations is developed, considering the restrictions of the operation and the predictions of active power production of the wind farms.

4.1. POWER CURTAILMENTS OF WIND POWER PRODUCTION

In the Spanish electrical system, some generators can inject all the available production to the system in normal situations, independently of the hourly market price. This is the Non-Dispatched Generation (NDG) of the system. Wind power, solar, hydraulic without regulation dam and other producers that use fluent or variable renewable energy are included in the NDG. However, the SO may require a reduction in the production of the NDG, if it is necessary to prevent risks in the system operation. This decrease can be demanded to the wind parks if there are not other means to avoid the risks. The following problems could be cause of NDG reductions [32]:

- Congestions in the generation provision: the SO calculates the maximum input of NDG that the system can absorb, in situations when the power predicted by the NDG can not be injected into the system, either because of excessive wind power production or due to restrictions in the transmission grid.
- Stability reasons: wind parks are sensitive to temporary voltage reductions (voltage dips). Reductions in the voltage value (even for periods lower than 100 ms.) could produce disconnections of older technology wind parks, with the subsequent stability risks. The SO must calculate the maximum of NDG that could be instantaneously disconnected by a voltage dip in each bus of the system, without causing stability problems. In this calculation, the SO must consider the sensitivity of each wind park to the voltage dips.
- Viability of the power system balances: the SO must guarantee both active and reactive power balances in the system. To perform this task, technical restrictions can be imposed by the SO to dispatched and not-dispatched generation units. The SO must consider possible generation surpluses, produced by either generations greater than forecast or demands lower than the expected loads.

Based on security analyses, the SO calculates at every moment the maximum amount of NDG that can be safely injected to the system, for each transmission bus of the grid. When a reduction of the NDG is required in some part of the system, the SO must select which producers must decrease their production. This assignment involves numerous tasks: the SO requires accurate predictions of the NDG for the following periods, the producers send the information to the SO, which collect the data, the SO calculates the required reductions in each transmission node and finally the SO informs the producers of the generation maximums. These tasks must be performed generally in critical situations, increasing the effort of the SO in these circumstances. Therefore, some SO (as REE) had proposed the creation of regional DD's, to act in such critical situations. When the SO detects the need of a NDG reduction, communicates to the related DD the maximum of acceptable NDG for the transmission buses involved. The DD calculates the optimal reduction of the generation in the transmission bus, informing the producers of the new operational points. DD's operate as intermediate entities between the SO and the non-dispatched producers, acting (as specified in the current legislation) only in critical situations. In Spain, the largest amount (54,6%, [28]) of the NDG corresponds to wind generation. Therefore, in a first step, DD's only could manage wind power producers, being expected the incorporation of other non-dispatchable producers in the next years. The first Spanish DD's are (in 2007) being tested and it is expected that they will be in operation along 2008.

Under the present regulation [32], NDG producers may choose the way in which the reductions are calculated: either they follow the settings set by the SO, or they associate in a DD and follow their settings. In the last case, the producer must declare its wish to participate to the DD, signing an agreement. On the other hand, the SO calculates the required reduction by following a proportional rule (explained in the following sections of the chapter) while DD's can use a proportional rule or any other to find the optimal reduction for each of its members. In any case, the DD calculation methodology must be previously approved by the SO, and the operational constraints (including the maximum admissible NDG in the transmission node) must be satisfied. In the next sections, an alternative to the proportional rule that can be used by the DD is described, considering the different controllability of the producers.

4.2. RATE OF DECREASE FOR NON DISPATCHABLE GENERATION

Once the maximum of admissible NDG in the transmission buses has been determined, the SO matches these values with the predicted power injection in the nodes. Both, admissible NDG and forecasted production should correspond to the same period of the next future. Due to technological restrictions, there is a delay between the request of reduction and the effective application of the reduction. In the present analysis, the delay is 15 min. Therefore, the first period of application of the restriction will be (at least) 15 min. after the calculation.

In the transmission buses in which NDG prediction exceeds the admissible injection, a reduction of the NDG is necessary. The SO calculates the reduction of the NGD production bus by bus, using a proportional rule. Suppose that in a bus the prediction for the total production of the wind farms ($\sum_{j=1}^{ng} P_{Gj}^{Av}$, with P_{Gj}^{Av} the estimation for the available active power production of the wind farm j and ng the number of wind farms connected to the transmission bus) in a period of the future exceeds the admissible NDG (AdP_{Out}^{Max}) in the bus for the same period, as shown in (7).

$$\sum_{j=1}^{ng} P_{Gj}^{Av} > AdP_{Out}^{Max} \tag{7}$$

The rate of decrease (*DR*) can be calculated as (8).

$$DR = \frac{AdP_{Out}^{Max}}{\sum_{j=1}^{ng} P_{Gj}^{Av}} = \begin{cases} \text{if } 0.0 \leq DR < 1.0 & \text{NDG reductions} \\ \text{if } DR \geq 1.0 & \text{normal operation} \end{cases} \tag{8}$$

In (8), when *DR* varies between 0 and 1 the SO may require reductions in the active power production of the NDG.

The active power production of producer *j* can be calculated, using DR, as in (9).

$$P_{Gj} = DR \cdot P_{Gj}^{Av} = \frac{AdP_{Out}^{Max}}{\sum_{j=1}^{ng} P_{Gj}^{Av}} \cdot P_{Gj}^{Av} \tag{9}$$

4.3. CALCULATION OF POWER REDUCTION BY THE DELEGATED DISPATCHES.

When the SO communicates the possible existence of critical situations to the involved DD, implying probable NDG diminutions, the DD must calculate the reductions needed for not exceeding AdP_{Out}^{Max}. For this task, the DD may also use a proportional rule, as previously presented, or any other. However, the calculations performed by the DD and SO may differ in three points.

a) As presented in previous sections, a reasonably accurate prediction can be obtained for the average production of the wind farms for the next 15 min. In this period, the wind power prediction can be considered a known quantity. However, if the security of the system is evaluated for periods beyond 15 min, uncertainties associated with wind power previsions could be considered. In this aspect, DD calculations may have a relative advantage in relation to the prevision performed by the SO. Better communications with the producers, a better knowledge of the wind characteristics of the region and the possibility of relating the prediction of a wind farm to the others suggest that the DD forecasts could be more accurate than the SO predictions.

b) The proportional rate calculated in (8) does not consider the power losses in the internal grid to the DD. The wind farms are connected to the transmission node through a distribution network, supervised by the DD. Therefore, the total production injected into the transmission node is the added production of the generators, decreased by the losses in the DD internal grid. If the amount of active power losses in this network is small, (8) is a good approximation. However, if active power losses represent an appreciable value, the wind productions would be decreased more than required, causing economic losses to the producers because the injected wind powers would be below the admissible NDG. An improved proportional rate calculation, that allows increasing the NDG production to compensate the DD internal losses, is analyzed in [4].

c) In the DD calculation, the different technological control possibilities of the wind farms may be considered. When DR is calculated by the DD, the sharing of the power decrease can be performed only by the controllable wind farms. As presented in the next section, not all the wind parks can control the active power generation continuously. The ability of the wind farms to modify the production as wished depends on the technology implemented in the wind park. Suppose that only $m1$ wind farms, of the m generators belonging to the DD ($m1 \leq m \leq ng$), can modify voluntarily the active power generation in the DD. Also, suppose that a reduction of the NDG is required in the bus ($\sum_{j=1}^{ng} P_{Gj}^{Av} > AdP_{Out}^{Max}$). Therefore, the Decrease Rate considering only $m1$ controllable generators (DR_{m1}) is presented in (10), where AdP_{DD}^{Max} is the maximum admissible input of DD in the transmission bus.

$$DR_{m1} = \frac{AdP_{DD}^{Max}}{\sum_{j=1}^{m1} P_{Gj}^{Av}} = \frac{DR \cdot \sum_{j=1}^{m} P_{Gj}^{Av}}{\sum_{j=1}^{m1} P_{Gj}^{Av}} \tag{10}$$

The reduction effort is only assumed by the $m1$ controllable generators, as shown in (11).

$$P_{Gj} = \begin{cases} DR_{m1} \cdot P_{Gj}^{Av} = \dfrac{AdP_{DD}^{Max}}{\sum_{j=1}^{m1} P_{Gj}^{Av}} \cdot P_{Gj}^{Av} & j = 1, \ldots, m1 \\ P_{Gj}^{Av} & j = 1, \ldots, (m-m1) \end{cases} \tag{11}$$

In (11), when the required reduction of the production ($\sum_{j=1}^{m} P_{Gj}^{Av} - AdP_{DD}^{Max}$) exceeds the control capacity of the DD $\left(\left(\sum_{j=1}^{m} P_{Gj}^{Av} - AdP_{DD}^{Max}\right) > \sum_{j=1}^{m1} P_{Gj}^{Av}\right)$, the non-controllable generators of the DD may be called on to switch off the complete production, to satisfy the SO constraints.

The ($m-m1$) non-controllable units will compensate the controllable wind farms for the reduction service, after the operation. Many payment schemes can be used for this compensation, for instance, to share it proportionally among the non-controllable generators, or according to the rated power, to the production, or considering controllability prices, etc. The consideration of controllability (and interruptive) prices in the payment scheme allows to implement a method based on market rules within the DD, as shown in next sections. Firstly, a classification of the wind power generation units, based on their controllability and interruptive abilities, is presented.

4.4. CLASSIFICATION OF WIND POWER PRODUCERS

Three main technologies dominate now the wind turbine market: squirrel cage induction generators directly coupled to the grid, direct drive synchronous generators and doubly fed induction generators [31]. The controllability of the wind farm turbines depends on the technology used. Doubly fed induction generators may vary continuously their production within their operational limits. This kind of turbines can follow different control strategies, as shown in [1], [8] and [9].

However, wind producers able to control their production could decide not to participate in the DD reduction service. Their willingness for active participation in the DD may depend on the benefits of participating. Therefore, the controllable wind power producer can offer its production control to the other producers, putting a price to this service, the Controllability Price.

All the wind farms have the capacity to cut off the complete production, disconnecting the wind farm from the network. In fact, this is the basic measure that the SO applies to the wind parks, switching off the parks in the connection bus under contingences. The wind farm owners can choose between allowing the DD (or the SO) to disconnect them only in very critical situation, or to put also a price to this service, the Interruptive Price. Partial curtailments (disconnecting only some of the wind park turbines) could be also possible, but they have not been considered in the model. However, the consideration of this ability can be easily included in the formulation.

According to the controllability and the interruptive aptitudes of the wind farms, they can be classified into the following groups:

- Type 1 wind parks offer to the DD both interruptive and controllability prices, aiming to participate in the control of the DD active power production under contingencies. Their services will be used if the price they give for them is low enough. When the wind farm can not be controlled but the producer wants to participate in the reduction service, by offering its interruptive ability, he may submit a very high controllability price. In [5], it is shown that the control of the power factor is necessary to perform an adequate production control. Therefore, Type 1 wind farms can be required to set their production and their power

factor, between the limits imposed by the reactive power production of the generator and the available wind power. The reactive power control is not paid in the present approach. However, a payment for reactive power controllability might be included in the model without problems.
- Type 2 wind parks contribute to the operation of the DD by controlling the power factor of their production. As previously expressed, in the present paper the contribution of these producers to the system operation is not paid. The active power output of Type 2 wind farms is the available wind power at the moment.
- Type 3 wind parks are producers that cannot, or do not wish to control either their active power generation or their power factor. The active power delivered by these wind farms is the available wind power at each instant. The power factor of these wind farms is one fixed value, according to the existing incentives for power factor specifications [33].
- Type 4 wind parks are those WPG that do not belong to the DD, but that are connected to the same output transmission bus. Under current regulation, this situation is possible. Type 4 Wind Parks receive the reduction setting directly from the SO, who follows a proportional rule (8). Like Type 3 wind farms, the Type 4 wind power producers usually specify a fixed power factor for the production.

Wind parks of Types 2, 3 and 4 may be interrupted, in very critical situations and as last choice, to solve the SO constraints.

4.5. Optimal Allocation of Power Curtailments under Market Rules

In the power reduction imposed by the SO, all the producers are considered controllable and with equal controllability cost. Therefore, the reduction effort is equally distributed among the NDG, using a proportional factor. However, some of the producers can not be controlled (facing interruptions in critical situations) or do not want to control the production (because of the management and maintenance costs associated). Furthermore, the technological alternatives of production introduce different charges for the control actions. Therefore, the costs for both control and interruptive actions will vary among the producers. A methodology based in bids is analysed in the present section, considering that the Type 1 wind parks offer within the DD their controllability and interruptive abilities. The prices submitted by these producers for their controllability and interruptive actions must consider the following aspects:

- Cost of reducing the production: the producers that reduce the generation to reach SO constraints will decrease the revenue obtained for injecting active power generation into the system. The price of the NDG depends on the regulation adopted to reward this kind of generation. In general, to calculate the price of NGD market prices and possible governmental incentives must be considered.
- Management costs: to control the active power generation in a wind farm requires to monitor the production, to decrease the generation in some of the turbines and probably to switch off others, when necessary. Besides this, the set-points of the generators (or electronic devices) must be handled to reach the reactive power generation required for a secure operation. Specific software, control equipments and specialized personnel must be employed.
- Amortization of the controllability equipments and increased maintenance costs: possible increments in the maintenance expenses for additional operations and the amortization of the control devices must be considered in the prices offered for reduction operations.

To simplify the formulation, maximum ramps for increasing/decreasing generation between consecutive periods and start-up costs were not considered. These aspects are generally considered not significant in wind turbines. The next section develops the mathematical formulation proposed.

4.6. MATHEMATICAL FORMULATION

The mathematical formulation used to distribute optimally the reduction effort among the DD producers for the next hours, considering prices for controllability and interruptive services, is presented in equations (12) to (26).

$$\min \sum_{j=1}^{m1}\left(cp_j \cdot CR_j + ip_j \cdot IR_j \cdot P_{Gj}^{Av}\right) + \sum_{j=1}^{m2+m3} fnp \cdot \left(S_{Gj} \cdot \cos\varphi_j - P_{Gj}^{Av}\right)^2 + \ldots$$
$$\sum_{j=1}^{m3+m4} fnc \cdot \left(\varphi_j - \varphi_f\right)^2 + \sum_{j=1}^{m4} fnp \cdot \left(S_{Gj} \cdot \cos\varphi_j - DR \cdot P_{Gj}^{Av}\right)^2 \quad (12)$$

s.t. $P_{out} \leq AdP_{Out}^{Max}$ (13)

$S_{Gi} \cdot \cos\varphi_i - P_{Di} - P_i(V,\alpha) = 0$ (14)

$S_{Gi} \cdot \sin\varphi_i - Q_{Di} - Q_i(V,\alpha) = 0$ (15)

$\alpha_{sk} = 0$ (16)

$S_{Gj} \cdot \cos\varphi_j + CR_j = P_{Gj}^{av} \cdot (1 - IR_j)$ (17)

$\dfrac{CR_j}{P_{Gj}^{av}} \leq 1.0 - IR_j$ (18)

$\varphi_f = \begin{cases} \cos^{-1}(fpf) \text{ for capacitive } fpf \\ -\cos^{-1}(fpf) \text{ for inductive } fpf \end{cases}$ (19)

$DR = \dfrac{AdP_{Out}^{Max}}{\sum_{j=1}^{ng} P_{Gj}^{Av}}$ (20)

$S_{Gi} \cdot \cos\varphi_i \geq 0$ (21)

$\cos\varphi_j \geq \cos\varphi_j^{min}$ (22)

$V_i^{min} \leq V_i \leq V_i^{max}$ (23)

$-T_{ik}^{max} \leq T_{ik} \leq T_{ik}^{max} \quad i \neq k$ (24)

$CR_j \geq 0$ (25)

$IR_j = \{0;1\}$ (26)

$i, k = 1...n$

$j = 1...m1$

In equations (12)-(26), cp_j and CR_j are the controllability price and controllability reduction factor of wind producer j, respectively; ip_j is the price of wind producer j for disconnection; IR_j is a binary variable, representing the connection of wind producer j; fnp and fnc are the weight coefficients for maintaining the active power production (of Types 2, 3 and 4 wind farms) and the power factor (for Types 3 and 4 wind farms) in the specified values, respectively; DR is the proportional factor, used by Type 4 wind farms to calculate their required active power production; P_{out} is the total active power output of the DD area; P_{Di} and Q_{Di} are active and reactive power demands at bus i, respectively; V_i and α_i are the module and angle of the bus voltage, respectively, at bus i, α_{sk} is the voltage angle at the slack bus; $\cos \varphi_j^{min}$ is the minimum power factor of the j producer; V_i^{min} and V_i^{max} are minimum and maximum limits of the voltage module in bus i; T_{ik} is the apparent power transmission between buses i and k; T_{ik}^{max} is the maximum apparent power limit of the transmission line between buses i and k; fpf is the specified power factor in Types 3 and 4 wind farms; m $(m=m1+m2+m3)$ is the number of wind farms included in the DD action; $m1$, $m2$, $m3$ and $m4$ $(m4=ng-m)$ are the numbers of Types 1, 2, 3 and 4 producers; and n is the number of buses at the internal of the DD system, including the transmission node.

Objective function (12) minimizes the cost for reduction actions, both for interruptive and controllability procedures. This is the primordial objective of the DD action, to distribute the reduction effort among the more adequate units. Also, objective function (12) includes some equality restriction, by using large-value multipliers fnp and fnc. In Types 2 and 3 wind farms, the active power generation is not controlled; therefore the production must be equal to the available wind power in each instant. Type 2 and 3 wind farms generate at a fixed production angle, φ_j. Finally, the generation in Type 4 producers is specified by the SO, following a proportional rule. These equality equations are included in the objective function, instead as equality constraints, to improve the convergence of the algorithm in very critical situations. In addition, violations of these equations will show possible corrective alternatives in non factible solutions.

Restriction (13) includes output restriction specified by the SO for NDG in the transmission bus. The power flow nonlinear equations within the DD internal grid are shown in (14) and (15), considering a null value of the voltage angle at the slack bus (16). The discrete variable IR_j specifies the unit commitment for the j producer. If $IR_j = 1$, the j wind farm is disconnected; if $IR_j = 0$, the j wind park is connected. In connected wind farms, the available active power can be decreased in a continuous mode, modifying CR_j, as shown in (17). The continuous control CR_j can only be used in connected producers with $IR_j = 0$, as expressed in (18). The fixed production angle for Types 2 and 3 wind farms is calculated in (19), as function of the specified power factor *fpf*. The proportional factor *DR*, used by the SO in Type 4 wind farms, is obtained in (20), as specified in (8). Minimum values for the active power production and the power factors of the producers are shown in (21) and (22). Operational constraints for voltages and apparent power flowing in the transmission lines within the DD area must be also considered in the model, (23) and (24).

The optimization problem (12)-(26) must be executed for each programming interval of the future. The model results in a mixed-integer nonlinear optimization problem. It can be solved by using Branch and Bound, Branch and Cut [6], [7] and [3915], Lagrangian relaxation [393], [24], [37], [39] and [40] or other methods [26]. In the present case, an exhaustive enumeration algorithm was implemented [39], [26], due to the low number of discrete variables. It is reasonable to assume that up to 6 big wind farms may be injecting power to the same transmission node.

4.7. TEST CASE

The method was evaluated in a real network, extracted from the Spanish National grid. In this test case, 6 wind farms inject power into the system through a transmission node. The main characteristics of the internal grid and the wind farms are shown in Table 8 and Table 9.

Table 8. Internal DD System

From	To	R [pu]	X_L [pu]	B_C [pu]	T_{ik}^{max} [pu]
1	2	0.00132	0.0720	0.0040	14.2
2	3	0.00360	0.1845	0.0008	4.8
2	4	0.00420	0.2153	0.0001	4.4
2	9	0.00500	0.2563	0.0007	2.7
3	5	0.00120	0.0615	0.0010	2.2
3	6	0.00180	0.0692	0.0016	2.4
4	7	0.00160	0.0615	0.0014	2.0
4	8	0.00680	0.0307	0.0020	2.2

Table 9. Buses and Wind Farms

Bus	Producer Name	Type	P_{Gj}^{Max} [pu]	ip_j [€/MWh]	cp_j [€/MWh]	B [pu]
1	-	-	-	-	-	0.899
2	WG2	1	5.000	75	150	2.100
3	-	-	-	-	-	0.000
4	-	-	-	-	-	0.000
5	WG5	1	2.000	72	150	0.000
6	WG6	2	2.200	-	-	0.000
7	WG7	4	1.800	-	-	0.000
8	WG8	1	2.100	82	87	0.049
9	WG9	1	2.500	78	84	0.000

In Table 8, the pu values of the resistance (R), series reactance (X_L), parallel susceptance (B_C) and maximum apparent power flow (T_{ij}^{max}) in the transmission lines of the grid are shown. Apparent power and voltage bases are 10 MVA and 66 kV, respectively. In the simulations, the output and slack bus is 1 and the voltages vary between 0.95 and 1.05. As shown in

Table 9, there are 6 wind generators in the test system, with rated capacities P_{Gj}^{Max}. One of them (at bus 7) does not want to participate in the DD control, although it is injecting power to the same transmission node (bus 1). Another one (connected to bus 6) only controls reactive power, injecting all the available active wind power to the grid. The other four aim to participate in the reduction control, submitting both controllability and interruptive prices to the DD. The wind power producers at buses 2 and 5 cannot control the active power production, offering a large value for this service. It must be stressed that the prices presented in

Table 9 are only orientative values, used to show the operation within the DD. In this table, the compensation susceptance (B) in the buses is also included.

In Figure 25, the test system is represented graphically. Bus 1 is the transmission node, in which the wind farms inject the production into the system.

In the present formulation, the solution of optimization problem (12)-(26) provides the optimal generation profile for one period of 15 minutes. In the next simulations an interval of 8 hours is simulated, solving 32 consecutive optimization problems (one for each 15 minutes period). In Figure 26, the available and admissible wind powers for each period are shown.

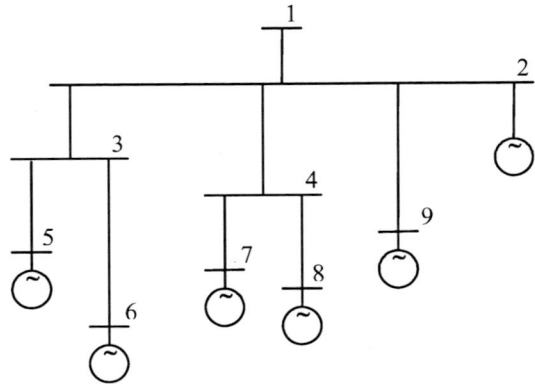

Figure 25. Test System.

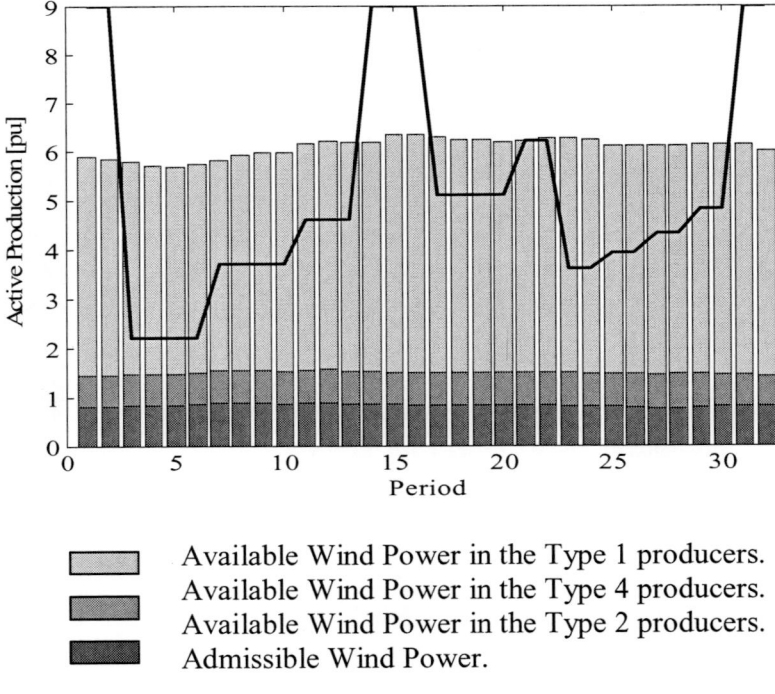

　　　　Available Wind Power in the Type 1 producers.
　　　　Available Wind Power in the Type 4 producers.
　　　　Available Wind Power in the Type 2 producers.
　　　　Admissible Wind Power.

Figure 26. Available and Admissible Wind Powers.

In Figure 26, the available wind power (classified by controllability type) and the output restriction imposed by the SO are depicted, in all the 32 programming periods. In periods 1-2, 14-16 and 31-32 there are not restrictions for injecting wind power into the transmission node. In all the others periods, the SO constraints the admissible wind power at different levels. The available wind power in Type 1 producers constitutes the large amount of renewable production, resulting of the addition for the previsions of WG2, WG5, WG8 and WG9. Also, in Figure 26 the available wind powers of Types 2 and 4 wind power producers are shown. There are not Type 3 wind power producers in the considered system (

Table 9). In Figure 27, the results obtained by solving the optimization problem in each period are shown.

As shown in Figure 27, SO constraints can be solved only by using the combined abilities of Types 1 and 4. If the admissible wind power is below the available production of the controllable producers (Types 1 and 4), disconnections of non-controllable Types 2 and 3 may be required.

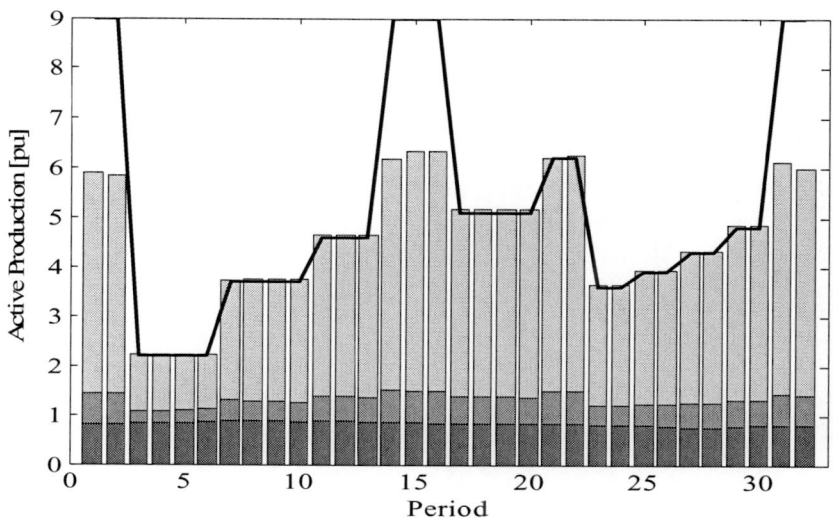

Figure 27. Production and Admissible Wind Powers.

In the presented simulations, the optimization problem (12)-(26) gives the amount of generation in Type 4 producers following the SO proportional rule (8), it maintains constant (if possible) Type 2 active power production and determines the optimal production of the Type 1 wind power generators. In those periods when a decrease of the total wind power production is required, it can be observed (Figure 27) that the added production is slightly higher than the admissible wind power. The optimization problem explicitly considers the active power losses in the operation, increasing the Type 1 production to closely reach the admissible wind power limit at the transmission node.

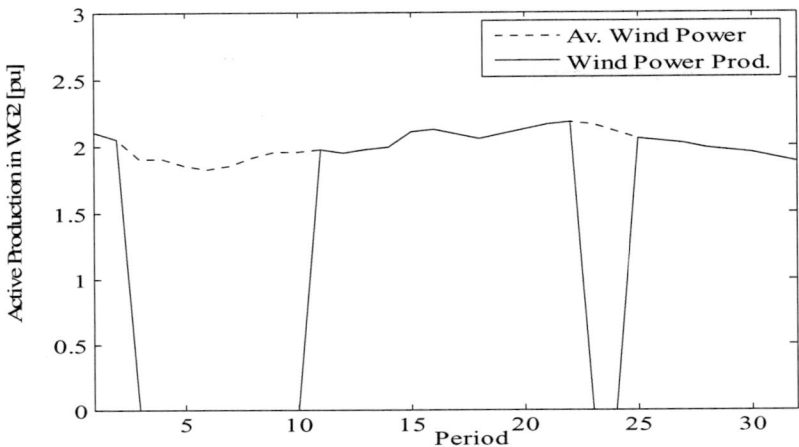

Figure 28. Production and Available Wind Power in WG2.

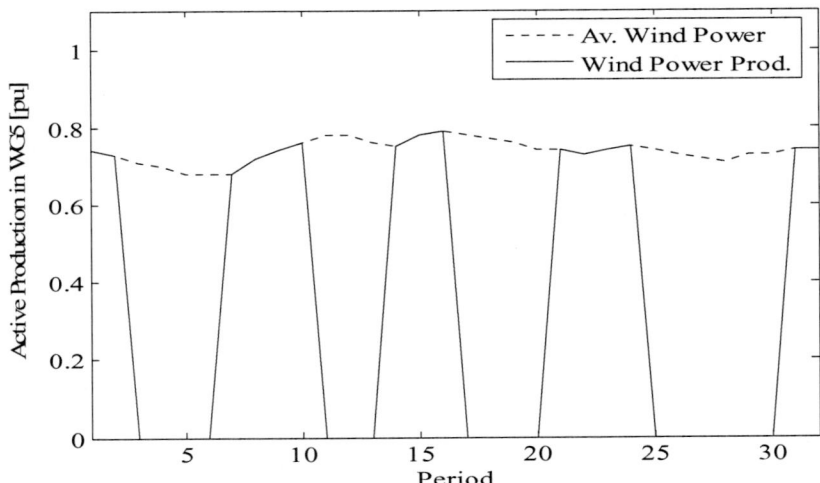

Figure 29. Production and Available Wind Power in WG5.

The best combination of generations in the Type 1 productions depends of both controllability and interruptive abilities of each producer, and the corresponding prices for these services. In Figure 28 to Figure 32, the production of the controllable producers is shown.

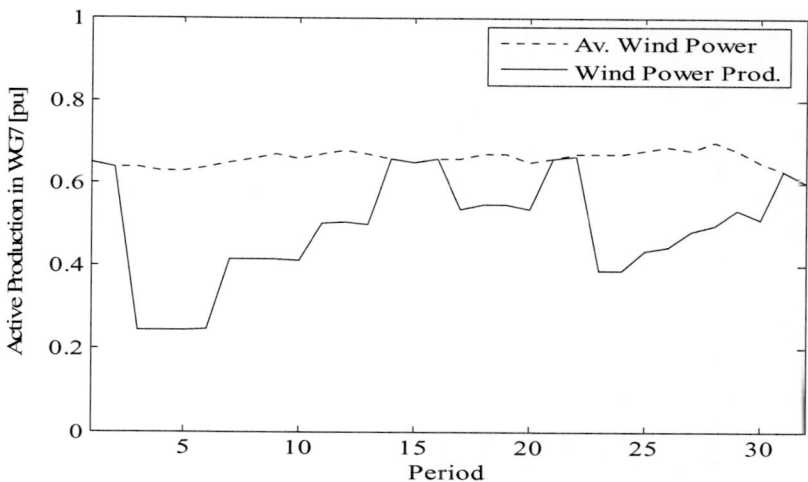

Figure 30. Production and Available Wind Power in WG7.

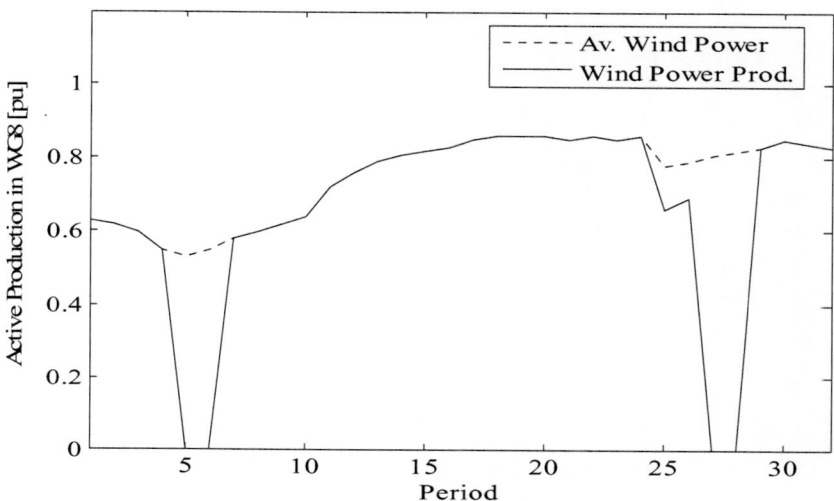

Figure 31. Production and Available Wind Power in WG8.

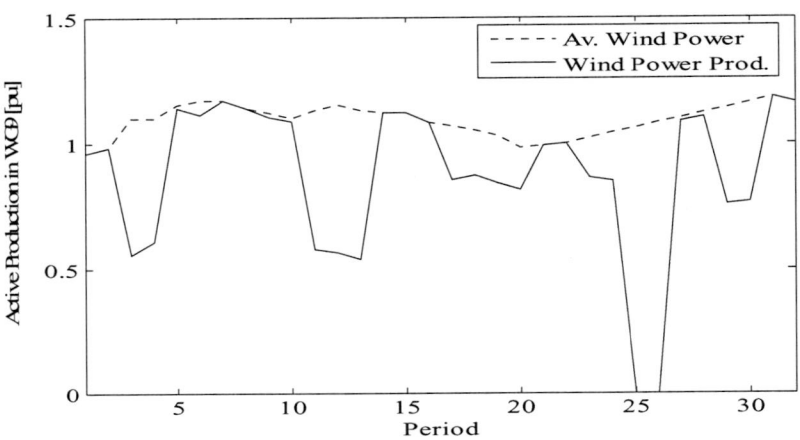

Figure 32. Production and Available Wind Power in WG9.

Two control strategies converge in transmission node 1. WG7 follows the proportional rule, as calculated by the SO. Therefore, this producer must modify continuously its active power generation (Figure 30) in all the periods in which restrictions on wind power are applied. The SO proportional rule (8) depends on the admissible wind power and on the wind power predictions for the wind farms connected to the same transmission node. Thus, the decrease rate DR varies in each interval and WG7 must adapt its own generation to this variation. The other controllable producers (WG2, WG5, WG8 and WG9) are managed by the DD. Wind power producers with only interruptible abilities (WG2 and WG5) are called on to switch off the generations when necessary. WG5 (Figure 29), with the lowest interruptible price, is firstly requested to suspend its generation. Larger wind power reductions demand the disconnection of WG2. Depending on the intensity of the reduction and the power generation, other controllable producers (WG8 in periods 5, 6, 27 and 28 and WG9 in periods 25 and 26) may be called to interrupt their production. The lowest controllability price was offered by WG9. Therefore, this producer is required to permanently vary the active power generation to follow, as close as possible, the maximum of available wind power accepted by the system at bus 1. Only when WG9 is called to interrupt the production (intervals 25 and 26), WG 8 (the other controllable producer) is requested to control the active power production. As expected, in the solution of the optimization problem the DD producers without continuous control abilities (WG2 and WG5) are not called on to control the production at any interval.

4.8. EX-POST ANALYSIS. ECONOMIC BALANCE

After the operation, the economic balance for all the participants of the DD must be settled. The producers that reduced their own generation to match SO requirements must be rewarded by the others. This operation is performed for each period. As an example, the 5th period is considered. In Table 10 and Table 11, the productions and flows in this period are synthesized.

In the 3th column of Table 10, the available wind power of the wind parks is shown. DD manages 5.05 pu of available wind power and 0.63 pu (of WG7) is directly controlled by the SO, totalizing 5.68 pu. This total amount of wind power can not be injected into the transmission grid, due to the imposed restrictions, and should be reduced to 2.2 pu. Following a proportional repartition, the SO calculates a decrease rate (8), as shown in (27).

Table 10. Active Power Balance at Interval 5. $ADP_{OUT}^{MAX} = 2.2$ **p.u.**

Management	Producer	P_{Gj}^{Av} [pu]	SO Proportional [pu]	Results of Optimization [pu]	DD Proportional [pu]	ΔP_j [pu]
SO	WG7	0.6300	0.2440	0.2440	--	0.0000
DD	WG2	1.8500	0.7165	0.0000	0.7240	0.7240
	WG5	0.6800	0.2634	0.0000	0.2661	0.2661
	WG6	0.8400	0.3254	0.8400	0.3287	-0.5113
	WG8	0.5300	0.2053	0.0000	0.2074	0.2074
	WG9	1.1500	0.4454	1.1363	0.4500	-0.6863
	Total	5.0500	1.9560	1.9763	1.9763	0.0000
Total		5.6800	2.2000	2.2203	--	0.0000

$$DR_{SO} = \frac{AdP_{Out}^{Max}}{\sum_{j=1}^{6} P_{Gj}^{Av}} = \frac{2.2}{5.68} = 0.3873 \qquad (27)$$

Applying DR to the available wind powers, results the SO proportional reduction shown in 4th column of Table 10. WG7, controlled by the SO, produces (0.3873·0.63=0.244 pu) in this period. To match the SO output restriction, the total DD injection should be (0.3873·5.05=1.956 pu). The solution of optimization problem (12)-(26) (5th column of Table 10) explicitly considers the losses, therefore the amount generated within the DD is slightly greater, 1.9763 pu (1% of increased production).

Table 11. Economic Balance at Interval 5. $\lambda = 71$ €/MWh, $\lambda_{DD} = 84$ €/MWh

Management	Producer	SO Proportional [€]	Results of Optimization [€]	Reduction Control [€]	Total Remuneration [€]
SO	WG7	43.3100	43.3100	--	43.3100
DD	WG2	127.1788	0.0000	152.0400	152.0400
	WG5	46.7535	0.0000	55.8820	55.8820
	WG6	57.7585	149.1000	-107.3630	41.7370
	WG8	36.4585	0.0000	43.5540	43.5540
	WG9	79.0585	201.6933	-144.1130	57.5803
	Total	347.1900	350.7933	0.0000	350.7933
Total		390.5000	394.1033	--	394.1033

After the operation, the effective contribution of the wind farms to the reduction effort is calculated. Considering only DD producers and the real generation, the proportional decrease rate within the DD is obtained in (28).

$$DR_{DD} = \frac{AdP_{DD}^{Max}}{\sum_{j=1}^{5} P_{Gj}^{Av}} = \frac{1.9763}{5.05} = 0.3913 \qquad (28)$$

In the 6th column of Table 10, the requested reductions (following the DD proportional rule, (28)) for the DD producers is shown. As shown in the 5th column of this table, the economic rule used in the optimization problem implies that some producers have increased reductions and other ones inject all their available wind power. The difference between the productions calculated using either DD proportional or economical rules is shown in 7th column of Table 10.

The internal price for the DD reduction effort is determined by the highest price unit required to reduce the production. In the present case, the internal DD reduction price is specified by the controllability of WG9, $\lambda_{DD} = 84$ [€/MWh]. Therefore, WG6 should pay to the DD producers that reduced the production in an amount equal to (84·0.5113·10·0.25 = 107.36 €). In the 5th column of table 11, the payments and rewards for the reduction effort within the DD producers is represented. Adding 3th and 5th columns of the table, the total remuneration of the producers is shown in the 6th column. In synthesis, the formula for calculating the remuneration of a wind park acting in the DD is presented in (29).

$$Profit_j = \lambda \cdot S_{Gj} \cdot \cos\varphi_j + \lambda_{DD} \cdot \left(DR_{DD} \cdot P_{Gj}^{Av} - S_{Gj} \cdot \cos\varphi_j\right) \qquad (29)$$

Chapter 5

CONCLUSION

The high penetration of this resource in the grids has raised concerns about the best way of integrating it into the power networks. The nature of wind power implies that the power production is highly changing and that it is difficult to manage with the traditional power system analysis and control tools. This leads to the need of developing new ones and to implement operation strategies that facilitate the integration of this clean and renewable resource without decreasing the efficiency and reliability of power grids.

In the previous sections, some of these new tools have been shown. The short term wind power prediction tools have been demonstrated to be necessary for the power system operation in grids with high wind power penetration. They have a reasonable accuracy, and are easy to implement and include in a power system control centre. When these predictions are produced for several wind farms, the accuracy of the joint prediction is even greater, and if predictions for only a wind farm (or a small number of them) are needed, then a higher detail in the wind farm modelling may lead to an assumable accuracy. This is still an open field for research, since more intensive computation tools (for numerical weather predictions) will lead to higher accuracies, and also new statistical and mesoscale models will give better results, according to the promising results of the research in this subject.

The integration of wind energy is also helped if wind generation participates in electricity markets as any other kind of generation. This leads to a greater implication of wind power producers into the system management and operation, and drive them to minimize the imbalance between the committed and the generated powers. Of course, prediction programs play a basic role in this participation and this minimization of deviations. It has been shown that these imbalances, when a prediction program is used, have an assumable cost and that trading in intraday markets and taking into account the uncertainty of the prediction still reduces them to a lower value.

The need of keeping power system in a secure operation may lead to wind power curtailments at some moments. To decide which wind farms must reduce their production should be decided taking into account the technical possibilities of the wind farms, and the price that the farm owners give to this control. This is why an optimization process that uses market mechanisms is an adequate and transparent way of allocating power curtailments among a group of users. The consequences of following this strategy have been shown in the previous sections and they are reasonable and easy to accept and understand.

New forms of generation demand a new management of the grid, if we want to keep the reliability of power networks and to include the renewable sources of energy that are required by the society.

REFERENCES

[1] Almeida R.G., Castronuovo E. D., Peças Lopes J.A., "Optimum generation control in Wind Parks when carrying out system operator requests", *IEEE Trans. on Power Systems*, pp. 718-725, vol. 21, n. 2, 2006.

[2] Asociación Empresarial Eólica (2006) *Ejercicio de Predicción. Informe final*. Available at http://www.aeeolica.es.

[3] Bailey B., Brower M., Zack J., "Wind Forecast: Development and Application of a Mesoscale Model". *Wind Forecasting Techniques: 33 Meeting of Experts. Technical Report from the Internatinal Energy Agency. R&D Wind*. Ed. S.-E. Thor, FFA, Sweden, pp. 93-116. 2000.

[4] Castronuovo E.D., Usaola J., "Alternatives of revenue for corrective actions of wind generators in a Delegated Dispatch", Proc. of the *IEEE International Conference on Clean Electrical Power*, Capri, Italy, pp. 567-573, May 2006.

[5] Castronuovo E.D., Usaola J., "Optimal controllability of wind generators in a Delegated Dispatch", *Electr. Power Syst. Res*, vol 77, n. 10, pp. 1442 – 1448, Aug. 2007.

[6] Chen X., Bushnell M. L., *Efficient branch and bound search with application to computer-aided design*, Boston, USA, Kluwer Academic, 1996.

[7] Dillon T.S., Edwin K.W., Kochs H.D., Taud R.J., "Integer programming approach to the problem of optimal unit commitment with probabilistic reserve determination", *IEEE Trans. Power App. Syst.*, vol. PAS-97, pp. 2154-2166, Nov./Dec. 1978.

[8] Ekanayake J., Holdsworth L., Jenkins, N. "Control of DFIG wind turbines", *Power Engineering Journal*, vol. 17, n.1, pp. 28-32, 2003.

[9] Ekanayake J., Jenkins N., "Comparison of the response of doubly fed and fixed-speed induction generator wind turbines to changes in network frequency", *IEEE Trans. on Energy Conversion*, vol. 19, n. 4, pp. 800-802, 2004.

[10] Ensslin C., Ernst B., Rohrig K., Schlögl F. "Online Monitoring and Prediction of Wind Power in German Transmission System Operation Centers". *Proceedings of the EWEC'03*. Madrid (Spain), June 2003.

[11] Focken U., Lange M., Waldl H.-P. "Previento- A wind Power Prediction System with an Innovative Upscaling Algorithm. " *Proceedings of the EWEC'01*. Copenhagen (Denmark), July 2001.

[12] Giebel G. *The State-of-the-Art in Short-Term Prediction of Wind Power. A Literature Overview*. Available at http://anemos.cma.fr

[13] Giebel G. (ed.) *Wind Power Prediction using Ensembles*. Report Risø-R-1527(EN). Risoe, Roskilde, 2005.

[14] González G., Díaz-Guerra B., Soto F., López S., Sánchez I., Usaola J., Alonso M., Lobo M.G. "SIPREÓLICO- Wind power prediction tool for the Spanish peninsular power system." *Proceedings of the 2004 CIGRÉ 40th General Session & Exhibition*. París (France), August 2004.

[15] Huang K.Y., Yang H.T., Huang C.L., "A new thermal unit commitment approach using constraint logic programming", *IEEE Trans. on Power Systems*, Vol. 13, pp. 936-945, Aug. 1998.

[16] Kariniotakis, G., et al., "Next Generation Short-term Forecasting of Wind Power - Overview of the Anemos Project", *Proceedings of the EWEC'06*, Athens (Greece), March 2006.

[17] Kariniotakis G. et al. "What performance can be expected by short term wind power prediction models depending on site characteristics?". *Proceedings of EWEC'04*. London (England), November 2004.

[18] Kariniotakis G.N.; Pinson P. "Uncertainty of short-term wind power forecasts a methodology for on-line assessment." *Proceedings of the International Conference on Probabilistic Methods Applied to Power Systems*. Ames (USA), September 2004.

[19] Landberg, L.: *Short-term Prediction of Local Wind Conditions*. PhD-Thesis, Risø-R-702(EN), Risø National Laboratory, Roskilde, Denmark 1994, ISBN 87-550-1916-1.

[20] Lai S.Y., Baldick R., "Unit commitment with ramp multipliers", IEEE Trans. on Power Systems, vol. 14, n. 1, pp 58-64, Feb 1999.

[21] Martí Perez, I.: "Wind Forecasting Activities". *Proceedings of the First IEA Joint Action Symposium on Wind Forecasting Techniques*, Norrköping,

Sweden, December 2002, pp. 11-20. Published by FOI - Swedish Defence Research Agency.

[22] Martí I., Usaola J., Sánchez I., Navarro J., Roldán A., González G., Díaz-Guerra B. "Wind power prediction in complex terrain. LocalPred and SIPREÓLICO." *Proceedings of the 2003 European Wind Energy Conference & Exhibition.* Madrid (Spain). June 2003.

[23] Nielsen T.S., Madsen H., Tøfting J. "WPPT: A Tool for On-line Wind Power Prediction". *Wind Forecasting Techniques: 33 Meeting of Experts. Technical Report from the International Energy Agency. R&D Wind.* Ed. S.-E. Thor, FFA, Sweden,pp. 93-116. 2000.

[24] Nowak I., *Relaxation and Decomposition Methods for Mixed Integer Nonlinear Programming*, Birkhäuser Verlag, Switzerland, 2005

[25] OFGEM, 2000. *An Overview of the New Electricity Trading Arrangements.* Available at http://www.ofgem.gov.uk

[26] Padhy N.P., "Unit commitment – A bibliographical survey", IEEE Trans. on Power Systems, vol. 19, n. 2, pp 1196-1205, May 2004.

[27] Pinson, P., Siebert, N., Kariniotakis, G., "Forecasting of Regional Wind Generation by a Dynamic Fuzzy-Neural Networks Based Upscaling Approach.", CD-Rom Proceedings of the European Wind Energy Conference & Exhibition EWEC 2003, Madrid, Spain, June 16-19, 2003.

[28] Red Eléctrica de España, *El Sistema Eléctrico Español, avance del Informe 2006*, available on-line in www.ree.es .

[29] Sánchez I., Usaola J., Ravelo O., Velasco C., Domínguez J., Lobo M. G. "SIPREÓLICO- a wind power prediction system based on fexible combination of dynamic models. Application to the Spanish power system." *Proceedings of the 2002 World Wind Energy Conference & Exhibition.* Berlin (Germany), July 2002.

[30] Sánchez I. "Short-term prediction of wind energy production", *International Journal of Forecasting* (2006), 22, 43-56.

[31] Slootweg J.G., Polinder H., Kling W.L., "Representing wind turbine electrical generating system in fundamental frequency simulations", *IEEE Trans. on Energy Conversion*, vol. 18, n. 4, pp. 516-524, Dec. 2003.

[32] Spanish Secretary of Energy, *Operational Procedure 3.7, Programming of the Non-Dispatched Renewable Generation* (in Spanish), BOE n. 254, pp. 37020-37022.

[33] Spanish Ministry of Energy, RD 661/2007 (in Spanish), BOE n. 126 (2007), pp. 22846-22886, available on-line in www.boe.es .

[34] Usaola J., Angarita J. "Benefits of short term wind power prediction programs for the integration of wind energy in electricity markets." *Proceedings of the EWEC'06.* Athens (Greece), March 2006.

[35] Usaola J., Angarita J. "Bidding wind energy under uncertainty." *Proceedings of the International Conference on Clean Electric Power. Renewable Energy Resources Impact.* Capri (Italy). May 2007.

[36] Usaola J., Ravelo O., Sánchez I., Velasco C., Domínguez J., Lobo M. G., González G. "SIPREÓLICO, a wind power prediction toll for the Spanish peninsular power system operation. *Proceedings of the 2002 Global Wind Power Conference.* Paris (France), April 2002.

[37] Wang C., Shahidehpour S.M., "Ramp-rate limits in unit commitment and economic dispatch incorporating rotor fatigue effect", IEEE Trans. on Power Systems, vol. 9, n. 3, pp. 1539-1545, Aug. 1994.

[38] Wind Power Observation, Spanish Wind Energy Association, available on-line at http://www.aeeolica.org/english/

[39] Wolsey L.A., *Integer Programming*, Wiley-Interscience Series, John Wiley and Sons, New York, USA, 1998.

[40] Zhuang F. and Galiana F.D., "Towards a more rigorous and practical unit commitment by Lagrange Relaxation", IEEE Trans. on Power Systems, vol.3, n. 2, pp 763-773, May 1988.

INDEX

A

accuracy, 4, 8, 11, 12, 13, 16, 17, 21, 29, 31, 65
algorithm, 6, 53, 54
alternative, 16, 45
alternatives, 50, 53
amortization, 51
assessment, 68
assignment, 45
assumptions, 25, 28, 30, 37, 38, 40
Athens, 68, 70
atmospheric pressure, 4
availability, 4, 24

B

benefits, 23, 24, 25, 49
bias, 11, 27, 30, 31, 40

C

classification, 48
closure, 25, 26, 29
clusters, vii
compensation, 48, 55
computation, 5, 7, 13, 21, 65
constraints, 13, 17, 45, 48, 50, 51, 53, 54, 57
consumers, 23

contingency, 43, 44
control, vii, 1, 2, 43, 44, 47, 48, 49, 50, 51, 54, 55, 60, 65, 66, 67
convergence, 53
Copenhagen, 68
correlation, 8
costs, 1, 2, 28, 30, 35, 50, 51
customers, 3, 24

D

decisions, 17
deficit, 23
demand, 17, 60, 66
Denmark, 18, 20, 68
density, 31
deviation, 11, 12, 26
diamonds, 38
discrete variable, 54
distribution, 11, 14, 32, 35, 43, 47

E

economic losses, vii, 2, 47
electric energy, 1
electric power, vii
electrical system, 44
electricity, vii, 1, 2, 3, 23, 24, 25, 34, 36, 65, 70

energy, vii, 1, 2, 3, 4, 23, 24, 26, 30, 31, 34, 35, 38, 65, 66, 69, 70
England, 68
environmental regulations, 43
equality, 53
estimating, 3
Europe, 20
evacuation, 17

F

family, 8
farm efficiency, 7
farms, 2, 3, 4, 6, 8, 13, 14, 17, 20, 43, 46, 47, 50, 53, 54, 66
fatigue, 70
filters, 20
fluctuations, 1
forecasting, 20, 28
France, 68, 70
frequency distribution, 32

G

generation, 1, 2, 3, 17, 28, 41, 43, 44, 45, 48, 51, 53, 55, 57, 60, 61, 62, 63, 65, 66, 67
Germany, 20, 69
Greece, 68, 70
grids, 1, 3, 21, 65
groups, 13, 17, 49
growth, 13

I

images, 20
imbalances, vii, 3, 27, 30, 34, 38, 66
in situ, 44
incentives, 50, 51, 63
income, 4, 28, 30
independent variable, 28
induction, 49, 68
inertia, 15
injections, 43
insight, 13, 40

integration, 4, 65, 70
intensity, 60
interval, 31, 32, 35, 54, 55, 60, 63
Italy, 67, 70

L

legislation, 45
limitation, 17
literature, 5
logic programming, 68

M

management, 1, 2, 4, 50, 65, 66
market, vii, 2, 3, 23, 24, 25, 26, 27, 28, 29, 30, 35, 36, 37, 44, 48, 49, 51, 62, 66
market prices, 27, 28, 29, 37, 51
markets, 1, 2, 3, 17, 23, 24, 26, 27, 28, 30, 31, 34, 35, 37, 38, 40, 65, 70
measures, 8, 10, 16
models, 4, 5, 6, 7, 8, 14, 15, 16, 18, 20, 21, 65, 69
modules, 8

N

network, 47, 49, 54, 68
neural network, 8
neural networks, 8
New York, iv, 70
nodes, 43, 46

O

operator, 17, 67
operators, vii, 3, 34
optimal performance, 8
optimization, vii, 54, 55, 57, 60, 61, 62, 63, 66
organization, 5, 24

P

parameter, 16, 28, 29
Paris, 70
penalties, 27, 31
performance, 2, 4, 6, 8, 16, 40, 68
portfolio, 12, 13, 17
power, vii, 1, 2, 3, 4, 5, 6, 7, 8, 9, 11, 12, 13, 14, 15, 17, 18, 21, 23, 24, 25, 28, 29, 31, 32, 34, 35, 36, 37, 38, 40, 43, 44, 45, 46, 47, 48, 49, 50, 51, 53, 54, 55, 57, 60, 61, 62, 65, 66, 68, 69, 70
power generation, 47, 49, 50, 51, 53, 60, 62
prediction, vii, 1, 2, 3, 4, 5, 6, 7, 8, 9, 10, 11, 12, 13, 15, 16, 17, 18, 20, 21, 23, 24, 25, 26, 27, 28, 29, 31, 32, 34, 35, 36, 37, 38, 40, 41, 46, 47, 65, 68, 69, 70
prediction models, 8, 68
predictors, 16
price taker, 28
prices, 28, 35, 36, 38, 40, 41, 48, 49, 51, 55, 59
probability, 15, 31, 32, 34, 35, 36
probability density function, 31, 32, 35
probability distribution, 15, 34
producers, vii, 2, 35, 43, 44, 45, 47, 49, 50, 51, 53, 54, 55, 57, 59, 60, 61, 62, 63, 65
production, 1, 2, 3, 4, 6, 7, 8, 9, 13, 14, 15, 18, 20, 24, 28, 31, 32, 34, 35, 36, 43, 44, 45, 46, 47, 48, 49, 50, 51, 53, 54, 55, 57, 59, 60, 62, 63, 65, 66, 69
profit, 6
program, 2, 3, 4, 5, 6, 8, 13, 18, 26, 28, 31, 34, 35, 66
programming, 54, 57, 67

R

radar, 20
range, 31, 32, 35
real time, 3, 4, 7, 14, 15, 16, 18, 31, 32
reduction, 25, 28, 30, 31, 44, 45, 46, 48, 49, 50, 51, 53, 55, 60, 61, 62, 63
regional, 5, 20, 24, 43, 45

regulation, 17, 44, 50, 51
regulatory framework, 25
relaxation, 54
reliability, 1, 65, 66
renewable energy, 44
reserves, 24
resistance, 55
resolution, 4, 5, 6, 14, 15, 17, 20, 21
revenue, 25, 26, 27, 29, 30, 31, 35, 40, 51, 67
rewards, 63
rolling, 28, 29
roughness, 4, 6, 11, 20

S

sample, 38
satellite, 20
scheduling, 4, 13, 17
search, 67
seasonal variations, 16
security, vii, 45, 47
sensitivity, 44
series, 38, 39, 55
shape, 31
sharing, 47
simulation, 20
skewness, 11
society, 66
software, 20, 51
Spain, 13, 14, 17, 20, 21, 43, 45, 68, 69
speed, 4, 5, 6, 15, 18, 68
spot market, 26
stability, 2, 44
standard deviation, 11
statistics, 8, 20
strategies, vii, 2, 49, 60, 65
subsidy, 2
Sweden, 67, 69
switching, 2, 49
Switzerland, 69
synthesis, 63
system analysis, 65
systems, vii, 2, 4, 38

T

tariff, 24
technology, 44, 47, 49
temperature, 4
threshold, 6
TID, 30
time, 3, 4, 6, 8, 9, 11, 13, 14, 15, 16, 18, 23, 24, 26, 27, 28, 29, 31, 32, 35
time lags, 32
time series, 6, 8, 14, 15, 18
total energy, 23
total product, 46, 47
trade, 23
trading, 23, 29, 30, 66
transmission, 43, 44, 45, 46, 47, 48, 50, 53, 54, 55, 57, 58, 60, 61
trend, 11, 21, 24, 38, 41

U

UK, 23
uncertainty, vii, 1, 2, 31, 34, 35, 36, 66, 70
updating, 4, 16, 26, 37
users, 18, 23, 66

V

values, vii, 1, 15, 27, 28, 29, 32, 37, 38, 40, 46, 53, 54, 55
variability, 1, 24
variable, 31, 44, 53
variables, 18
variance, 9, 10, 31, 32
variation, 60

W

wind, vii, 1, 2, 3, 4, 5, 6, 7, 8, 9, 10, 11, 12, 13, 14, 15, 16, 17, 18, 20, 21, 23, 24, 25, 26, 28, 29, 31, 32, 35, 36, 37, 38, 43, 44, 45, 46, 47, 48, 49, 50, 51, 53, 54, 55, 57, 60, 61, 62, 63, 65, 66, 67, 68, 69, 70
wind farm, vii, 2, 3, 4, 6, 7, 8, 9, 10, 11, 12, 13, 14, 15, 16, 17, 18, 20, 21, 24, 25, 26, 29, 31, 32, 35, 36, 38, 43, 44, 46, 47, 48, 49, 50, 51, 53, 54, 55, 60, 62, 65, 66
wind generators, vii, 1, 23, 24
wind turbines, 4, 6, 24, 51, 67, 68
windows, 28, 29

Y

yield, 34